The Bhutan Electric Vehicle Initiative

DIRECTIONS IN DEVELOPMENT
Infrastructure

The Bhutan Electric Vehicle Initiative

Scenarios, Implications, and Economic Impact

Da Zhu, Dominic Pasquale Patella, Roland Steinmetz, and
Pajnapa Peamsilpakulchorn

© 2016 International Bank for Reconstruction and Development / The World Bank
1818 H Street NW, Washington, DC 20433
Telephone: 202-473-1000; Internet: www.worldbank.org

Some rights reserved

1 2 3 4   19 18 17 16

This work is a product of the staff of The World Bank with external contributions. The findings, interpretations, and conclusions expressed in this work do not necessarily reflect the views of The World Bank, its Board of Executive Directors, or the governments they represent. The World Bank does not guarantee the accuracy of the data included in this work. The boundaries, colors, denominations, and other information shown on any map in this work do not imply any judgment on the part of The World Bank concerning the legal status of any territory or the endorsement or acceptance of such boundaries.

Nothing herein shall constitute or be considered to be a limitation upon or waiver of the privileges and immunities of The World Bank, all of which are specifically reserved.

Rights and Permissions

This work is available under the Creative Commons Attribution 3.0 IGO license (CC BY 3.0 IGO) http://creativecommons.org/licenses/by/3.0/igo. Under the Creative Commons Attribution license, you are free to copy, distribute, transmit, and adapt this work, including for commercial purposes, under the following conditions:

Attribution—Please cite the work as follows: Zhu, Da, Dominic Pasquale Patella, Roland Steinmetz, and Pajnapa Peamsilpakulchorn. 2016. *The Bhutan Electric Vehicle Initiative: Scenarios, Implications, and Economic Impact.* Directions in Development. Washington, DC: World Bank. doi:10.1596/978-1-4648-0741-1. License: Creative Commons Attribution CC BY 3.0 IGO

Translations—If you create a translation of this work, please add the following disclaimer along with the attribution: *This translation was not created by The World Bank and should not be considered an official World Bank translation. The World Bank shall not be liable for any content or error in this translation.*

Adaptations—If you create an adaptation of this work, please add the following disclaimer along with the attribution: *This is an adaptation of an original work by The World Bank. Views and opinions expressed in the adaptation are the sole responsibility of the author or authors of the adaptation and are not endorsed by The World Bank.*

Third-party content—The World Bank does not necessarily own each component of the content contained within the work. The World Bank therefore does not warrant that the use of any third-party–owned individual component or part contained in the work will not infringe on the rights of those third parties. The risk of claims resulting from such infringement rests solely with you. If you wish to reuse a component of the work, it is your responsibility to determine whether permission is needed for that reuse and to obtain permission from the copyright owner. Examples of components can include, but are not limited to, tables, figures, or images.

All queries on rights and licenses should be addressed to the Publishing and Knowledge Division, The World Bank, 1818 H Street NW, Washington, DC 20433, USA; fax: 202-522-2625; e-mail: pubrights@worldbank.org.

ISBN (paper): 978-1-4648-0741-1
ISBN (electronic): 978-1-4648-0755-8
DOI: 10.1596/978-1-4648-0741-1

Cover photo: © Roland Steinmetz. Used with the permission of Roland Steinmetz. Further permission required for reuse.
Cover design: Naylor Design.

Library of Congress Cataloging-in-Publication Data has been requested.

Contents

Preface	*xiii*
Acknowledgments	*xv*
Executive Summary	*xvii*
Abbreviations	*xxxiii*

Chapter 1	**Introduction**	1
	The Bhutan Electric Vehicle Initiative in Context	1
Chapter 2	**Background**	5
	Key Messages	5
	Bhutan's Macroeconomic Situation, Development Objectives, and Key Sectors	5
	Global EV Initiatives and the Context of the Bhutan EV Initiative	7
	Notes	8
	References	8
Chapter 3	**Scenarios for Electric Vehicle Uptake in Bhutan**	9
	Key Messages	9
	Global EV Penetration	9
	Influencing Factors for EV Adoption	10
	Potential Market Segments in Bhutan	13
	Three Scenarios for EV Uptake in Bhutan	16
	Notes	19
	References	19
Chapter 4	**Electric Vehicle Market and Technology Development**	21
	Key Messages	21
	Global EV Market Development	21
	Types of EVs: Plug-In Hybrids and Full Electric Vehicles	23
	Factors Influencing Driving Range	24
	An "Average Ride" in Bhutan	28
	International User Experience with EVs	29

	Battery Performance and Battery Second Life	32
	Notes	36
	References	37
Chapter 5	**Fiscal and Economic Incentives**	**39**
	Key Messages	39
	International Experience with Incentive Programs	39
	Analysis of Incentives and Total Cost of Ownership in Bhutan	42
	Notes	56
	References	56
Chapter 6	**Charging Infrastructure and Network Planning**	**57**
	Key Messages	57
	Importance of Charging Infrastructure	58
	Types of EV Charging and Available Standards	59
	EV Charging Options in Bhutan	63
	Charging Infrastructure Requirements by Uptake Scenario and National Rollout	81
	Market Models for Ownership and Operation of Charging Infrastructure	87
	Costs and Financing Arrangements	93
	Grid Impact and Power Quality	95
	Notes	105
	References	105
Chapter 7	**Policy and Economic Analysis**	**107**
	Key Messages	107
	Overview of the Policy and Economic Analysis	108
	EV Program Investment Requirements	108
	Fiscal Impact	110
	Fuel Import Benefits	113
	Impact on Trade Balance	116
	Environmental Benefits	121
	Note	124
	References	124
Chapter 8	**Stakeholders and Public Transport**	**125**
	Key Messages	125
	Stakeholder Analysis	125
	EV in a Broader Context: The Role of Public Transport in Green Mobility	133
	Note	135
	Reference	135

Appendix A	Background Information on Urban Transport in Bhutan	137
	Bus Services and Taxi Use	137
	Private Cars	137
Appendix B	Examples of International Incentive Programs	141
	China	141
	Japan	141
	Norway	142
	The United Kingdom	142
	The United States	143
Appendix C	Total Cost of Ownership Analysis for Bhutan	145
	Introduction to the Total Cost of Ownership Analysis	145
	Assumptions for the TCO Analysis	145
	Results of TCO Analysis for Private Vehicles	145
	Scenario Analysis for Setting Incentives—Private Vehicles	145
	Results of TCO Analysis for Taxis	152
	Results of TCO Analysis for Government Fleet	153
Appendix D	Suppliers of CHAdeMO (CCS/AC) Fast Charging Equipment	155
Appendix E	Possible Location of Charging Stations in Thimphu	157
Appendix F	Comparison of Bus Transport Technologies	161
	International Experience and Best Practice	161
	Bus Transport Total Cost of Ownership	163
	Life Cycle Analysis Carbon Dioxide Emissions	165
	Notes	167

Boxes

3.1	First Assessment: EV Awareness	12
3.2	First Assessment: Government Fleet as a Target Group	15
3.3	First Assessment: Target Group Private Vehicles	15
3.4	First Assessment: Taxi Vehicles as a Target Group	16
4.1	Case Study: Experiences with EV Taxis in the Netherlands	31
5.1	Norway's Fiscal and Economic Incentive Programs	44
5.2	TCO and Incentives in Other Countries	47
6.1	First Assessment: Charging Standards	63
6.2	First Assessment: Charging at Reserved Private Parking	68
6.3	First Assessment: Reserved Private Parking and Separate BPC Grid Connection	69
6.4	First Assessment: Reserved Public Parking with Extended Private Charging	71

6.5	General Requirements for Public Charging Stations	73
6.6	First Assessment: Public On-Street Parking and Slow Charging	74
6.7	Case Study: Public Charging in the Netherlands	74
6.8	First Assessment: Public Slow Charging	76
6.9	Technical Requirements for Fast Charging	77
6.10	First Assessment: Fast Charging	78
6.11	General Assumptions for the Calculation of Charging Infrastructure Requirements	82
6.12	Operations of a Charging Operator Back Office	90
6.13	First Assessment: Selecting a Charging Infrastructure Operator	92
6.14	Cost Assumptions for Charging Equipment	93
6.15	First Assessment: Grid Impact	97
6.16	First Assessment: Energy Use and Impact on the Grid	102
8.1	Who Are the EV Buyers in Bhutan?	129
8.2	Areas for Improvement of Thimphu Bus Services	135

Figures

3.1	EV Sales Volume and Share of Total Vehicle Sales in Top Ten of Leading EV Markets, 2012	10
3.2	Motivation for Electric Vehicle Purchases in California	13
3.3	Number of Electric Vehicles for the Three Uptake Scenarios, 2015–2020	19
4.1	Electric Vehicle Uptake Worldwide	22
4.2	Diffusion of Innovations Curve	22
4.3	Electric Vehicle Market Share per Country	23
4.4	Factors Influencing Electric Vehicle Power Consumption	25
4.5	Nissan Leaf: Range vs. Temperature	26
4.6	Tesla: Range vs. Speed	27
4.7	Nissan Leaf: Range Predictor	28
4.8	Elevation Profile of the Route Thimphu–Phuentsholing	28
4.9	Number of Journeys and Average Distance	29
4.10	Charging Locations Used in the Switch EV Trial	30
4.11	High-Level Composition of Battery	33
4.12	Battery Composition of a Lithium Iron Phosphate Battery (LFP Battery)	34
5.1	Fiscal Incentives and Electric Vehicle Penetration Rate, Selected Countries	41
5.2	Market Growth Rate vs. Per-Vehicle Incentive for Renault Zoe Battery Electric Vehicle, Private and Company Cars, 2012–2013	41
5.3	Disadvantages of an Electric Vehicle Identified by Different Groups of EV Owners in Norway	43
5.4	Combined Public Spending on Electric Vehicles in 15 Countries with EV Initiatives	43

5.5	Total Fiscal Support for Electric Vehicle Incentives in Norway	44
B5.1.1	Economic Incentives for Electric Vehicle Owners in Norway	45
B5.2.1	TCO for Mitsubishi i-MiEV (EV) vs. Fiat 500 (ICE), 2012 Pricing, Comparing Norway, Sweden, and Denmark	48
B5.2.2	Evaluation of TCO for France, Germany, and Norway	48
B5.2.3	Summary of TCO Calculations for Renault Clio vs. Renault Zoe (Private Car Market)	49
5.6	Savings for Private Vehicles when Switching to EVs, Annual Fuel Price Increase of 7 Percent at Different Discount Rates	51
5.7	Savings for Private Vehicles when Switching to EV, Annual Fuel Price Increase of 1 Percent at Different Discount Rates	52
5.8	Savings for Taxis using Various EV Options, Annual Fuel Price Increase of 7 Percent at Different Discount Rates	53
5.9	Savings for Taxis using Various EV Options, Annual Fuel Price Increase of 1 Percent at Different Discount Rates	54
6.1	Correlation between Number of Charging Stations and EV Uptake per Country	58
6.2	Illustration of Extended Private Charging in Public Space	69
6.3	Parties Involved in Public Charging	72
B6.7.1	Development of Public Investments in the Netherlands	75
6.4	Indicative Plan and Timetable for the Realization of the Charging Network	85
6.5	Integrated Infrastructure Model	87
6.6	Independent e-Mobility Model	88
B6.12.1	Screenshot of Back Office Charging Operator	90
6.7	Histogram of the Ratio between Expected Annual Peak Load of a Subnetwork and the Peak Load Limit of the Network	97
6.8	Typical Morning and Evening Peak Load on the Electricity Grid in Bhutan	98
6.9	Daily Electricity Use Pattern (Load Profile), 2010	98
6.10	Time and Distance Distribution for Average Travel	99
6.11	Typical Charging Profile at E-laad Public Charging Operator	99
6.12	DC Fast Charging Connection Time by Hour of the Day, by Region	100
6.13	Communication Structure for Smart EV Charging	104
6.14	Vehicle-to-Home Smart Grid	105
7.1	EV Purchases and Investment in Charging Infrastructure by Scenario, 2015–2020 (million Nu)	109
7.2	Private and Public Contribution by Scenario, 2015–2020 (million Nu)	109
7.3	Percentage Growth in Number of Vehicles, Fuel Imports, and Total Imports, 2003–2012	113
7.4	Fuel and Vehicle Imports in 2011	114
7.5	Fuel, Diesel, and Petro Imports by Value, 2009–2012 (% Growth)	114

7.6	Fuel, Diesel, and Petro Imports by Volume, 2009–2012 (% Growth)	115
7.7	Avoided Fuel Import by Scenario, 2015–2027	116
7.8	Bhutan's Balance of Payments	117
7.9	Incremental Impact on Imports, 2015–2027	118
7.10	Net Incremental Impact on Imports by Scenario in Value	119
7.11	Net Incremental Impact on Imports by Scenario as Percentage of Future Imports Projection	120
7.12	Net Incremental Impact on Imports for Private Vehicles, Taxis, and Government Fleet, High EV Uptake Scenario	120
7.13	Accumulated Avoided GHG Emissions by Scenario, 2015–2027	122
7.14	Accumulated Avoided GHG Emission from Private Vehicles, Taxis, and Government Fleet in High EV Uptake Scenario, 2015–2027	123
8.1	Share of Mean Monthly Per Capita Household Transport and Communication Expenses of Total Per Capita Household Nonfood Consumption by Quintile in Urban Area	132
A.1	Bhutan Urban Population	138
A.2	Motor Vehicle Ownership in Bhutan	138
A.3	Vehicle Registration in Bhutan, 1997–2013	139
C.1	TCO Comparison	151
C.2	Upfront Cost Comparison	151
C.3	TCO Comparison of ICE and EV for Taxis, Annual Fuel Price Increase at 7%	152
C.4	TCO Comparison of ICE and EV for Taxis, Annual Fuel Price Increase at 1%	153
F.1	Bus Transport Total Cost of Ownership per Kilometer for Different Technologies	165
F.2	Life Cycle Analysis Results: Annual CO_2 Emissions per Bus for Different Technologies	166

Maps

6.1	Draft BPC Plan for 46 Fast Chargers and Possible Phased Rollout	77
E.1	Possible Charging Locations in Thimphu Thromde	158

Photos

6.1	Examples of Charging Stations	62
6.2	Private Charging Station with Socket for EV Plug	65
6.3	Outdoor Charging Socket at Clearly Indicated EV Parking Space	66
6.4	Possible Private Parking Location at an Apartment	67
6.5	Example of Reserved Parking Place for EV Charging	68

6.6	Extended Private Charging Station with Two Sockets	71
6.7	Network of Charging Poles with 4 Sockets Connected to a Central Charging Hub	72
6.8	Examples of Taxi Stands Not Suitable for a Charging Location Because of Lack of Space	80
6.9	Example of Possible Charging Location for Taxis in Thimphu with Sufficient Space and Bhutan Power Corporation Connection	80

Tables

ES.1	Summary of Key Assumptions of Each EV Scenario and Estimates of Public Support Required during 2015–2020	xxv
ES.2	Summary of Policy and Fiscal Impact in Three Scenarios	xxix
3.1	Estimated Total Number of Vehicles in Bhutan, by Target Group	17
3.2	Number of Electric Vehicles by 2020, Low EV Uptake Scenario	18
3.3	Number of Electric Vehicles by 2020, High EV Uptake Scenario	18
3.4	Number of Electric Vehicles by 2020, Super High EV Uptake Scenario	18
4.1	Battery Capacity and Range of Selected Electric Vehicles	25
4.2	Original Equipment Manufacturer Battery Warranties	35
5.1	Financial and Nonfinancial Incentives	40
B5.1.1	Economic Incentives for Electric Vehicles and Resulting Costs for EVs vs. Petrol Cars in Norway	45
5.2	Current Fiscal Incentives for Electric Vehicles—Taxes on Selected Vehicles Used in the Analysis	50
5.3	Required Levels of Additional Cost Subsidy to Achieve High and Super High Uptake Scenarios	54
6.1	Types of EV Charging Based on Location and Type	59
6.2	Types of Charging and Charging Speeds	60
6.3	International Standards for Plugs and Sockets for Normal and Fast Charging	61
6.4	Charging Modes	62
6.5	Low EV Uptake Scenario: Estimated Number of Chargers Required by 2020	83
6.6	High EV Uptake Scenario: Estimated Number of Chargers Required by 2020	84
6.7	Super High EV Uptake Scenario: Estimated Number of Chargers Required by 2020	86
6.8	Advantages and Disadvantages of the Two Market Models	88
6.9	International Market Models	89
B6.14.1	Cost Assumptions for Charging Equipment	93

6.10	Indicative Cost per Scenario	94
6.11	Low EV Uptake Scenario: Total Energy Use and Maximum Peak Load in 2020	101
6.12	High EV Uptake Scenario: Total Energy Use and Maximum Peak Load in 2020	101
6.13	Super High EV Uptake Scenario: Total Energy Use and Maximum Peak Load in 2020	101
7.1	Overview of Private and Public Investments and Spending for the EV Initiative	108
7.2	Investment Requirement of EV Program in Three EV Scenarios, 2015–2020	109
7.3	Estimated Total Fiscal Impact from Incentives by Scenario, 2015–2020	111
7.4	Total Public Investment by Scenario, 2015–2020	111
7.5	External Trade in Million Nu	117
7.6	Total Impact of Imports from the EV Program, 2015–2020 (Million Nu and Percentage of 2012 Imports)	118
7.7	Avoided GHG Emissions by Scenario (tCO_2e)	122
7.8	Social Values of Avoided GHG Emissions in Different EV Scenarios	123
8.1	Initial Assessment of Stakeholder Impact	126
8.2	Estimated Annual and Monthly Fuel Expenses for ICE and Electric Vehicles	127
8.3	Summary of Potential Beneficiaries and Benefits	128
8.4	Distribution of Households That Own Transport Assets by Per Capita Consumption Quintile (Percentage)	130
8.5	Annual Household Income Distribution	130
8.6	Estimates of Household Affordability for EV Purchase	130
8.7	Mean Monthly Per Capita Household Food and Nonfood Expenditure by Per Capita Household Consumption Quintile in the Urban Area (Nu)	131
8.8	Estimates of Public Benefits and Costs	133
A.1	Composition of the Vehicle Fleet in Bhutan	139
B.1	EV Incentive Programs in China, India, Japan, the United Kingdom, and the United States (2014)	142
C.1	Assumptions for the TCO Analysis for Private Vehicles	146
C.2	Assumptions for the TCO Analysis for Taxis	147
C.3	Assumptions for the TCO Analysis for Government Vehicles	148
C.4	TCO and Switching Values—Private Vehicles	150
C.5	Cost Comparison of EV and ICE Vehicle (%) at 10% Discount Rate	151
D.1	List of Suppliers for CHAdeMO Fast Charging Equipment	155
F.1	Pros and Cons of Various Bus Technologies	163
F.2	Bus Transport TCO Assumptions and Results	164
F.3	Life Cycle Analysis Assumptions and Results	166

Preface

As the world is facing greater climate risk, there is a call for countries to take more timely actions to cut emissions. In the transport and urban sectors, low- or zero-emission technology and innovation that occur at a more rapid pace offer interesting potentials for countries to achieve economic and environmental objectives. While new technology adoption is traditionally viewed as a matter of private individual choice or consumer preference, today new technology and innovation are increasingly adopted in public policy agenda by governments to actively address sustainable development challenges.

The electric vehicle (EV) is one of the new technologies that has garnered attention in developed countries, and increasingly in developing countries, as one of the key green low-carbon urban mobility options. Although EV technology and market penetration are developing rapidly, EVs are still in a nascent stage. The technical, financial, and economic viability of EVs largely depends on various factors that are contingent on different contexts in which the technology is being introduced.

Bhutan is a country on a remarkable growth path. The country is also unique with its globally renowned commitment to a development philosophy that puts the importance of the environment on an equal footing with economic prosperity. Driven by hydropower development, electricity sales revenue, and a growing tourism sector, the country is growing at an impressive rate—between 7 and 8 percent for the past five years. Economic growth and rapid urbanization steadily increase demand for urban transport and imported fuels, but also further worsen the already large and growing trade and fuel deficit as well as raise concern over energy security. Addressing fuel dependency and promoting clean environment by increasing reliance on domestic clean hydropower is the country's development strategy.

In this context, the Royal Government of Bhutan (RGoB) has launched an ambitious initiative to promote the use of EVs to address both environmental and fossil fuel dependency. The RGoB thus requested the World Bank to provide a technical study to share global expertise and international experiences in planning and implementing EV initiatives in other countries. The World Bank welcomed this exciting opportunity to work with the Bhutanese government to conduct this study with the objective of informing the policy process and facilitating evidence-based policy debate that tends to be surrounded by technology uncertainties and lack of comprehensive information.

The study team worked across conventional technical boundaries and adopted an integrated approach by establishing links among urban development, transport, energy, environment, and climate change in the overall framework of the EV initiative. In addition, besides diving into the core technology issues of EVs, the study tries to address different areas that form an integral part of EV policy introduction—from stakeholder analysis and public awareness to economic, financial, and fiscal implication.

The study was a collaborative effort between the EV task force, led by the Gross National Happiness Commission, and the World Bank study team. We are most grateful for the opportunity to provide technical support to the government on EV policy development and share international experiences.

We highly appreciate the opportunity to work closely as well as exchange ideas and views openly with the Gross National Happiness Commission and all members of the RGoB's EV task force, which allowed the team to understand the unique policy and technical landscape of introducing EVs in Bhutan. The study team benefited greatly from the EV task force's guidance, knowledge, dedication, and proactive involvement throughout the process. Without such collaborative spirit, it would not be possible to conduct a study that is innovative, highly contextual, relevant, and responsive to our client's needs.

EV is an emerging area that is gaining more interest among policy makers and practitioners in other developing countries. We hope that the study will provide a strong technical background and policy options that will help the government make sound policy decisions in its EV initiative. We also hope that the analytical framework, knowledge, and experiences from Bhutan will benefit audiences working on similar policy agendas.

Acknowledgments

This report, *The Bhutan Electric Vehicle Initiative: Scenarios, Implications, and Economic Impact*, has been prepared under The World Bank Bhutan Green Transport and Electric Vehicle Initiative Technical Assistance Project by a Project Task Team consisting of Da Zhu (Task Team Leader), Dominic Pasquale Patella (co-Task Team Leader), Roland Steinmetz (electric vehicle [EV] consultant), Pajnapa Peamsilpakulchorn (economist, consultant), and Tenzin Lhaden, with contributions from Philip Sayeg and Paul Williams.

Royal Government of Bhutan EV Task Force, which comprises Lhaba Tshering (Gross National Happiness Commission), Nyingtob Pema Norbu (Gross National Happiness Commission), Sonam Dendup (Ministry of Information and Communications), Ujjwal Dahal (Bhutan Power Corporation), Pasang Tshering (Bhutan Postal Corporation), Ugyen Lhamo (Thimphu Thromde), Karma Pemba (The Road Safety and Transport Authority), Thinley Namgyal (The Road Safety and Transport Authority), and Rinchen Jamtsho (National Land Commission), provided technical inputs, comments, and relevant program data and information for the development of this report, as well as facilitated the study, discussion, and interview process. Early draft reports were reviewed and discussed with the task force members, and their time and valuable comments are highly appreciated.

The authors would like to thank Johannes Zutt (World Bank Country Director for Nepal, Bangladesh, and Bhutan), Robert J. Saum (former Country Director for Afghanistan and Bhutan), Karla Gonzalez Carvajal (Practice Manager, GTIDR), Ming Zhang (Practice Manager, GSURR), Sereen Juma (Country Program Coordinator, SACBN), and Genevieve Boyreau (Country Representative, Bhutan) for their guidance. The following peer reviewers contributed valuable comments: Stephen Alan Hammer, Genevieve Boyreau, Matias Herrera Dappe, Arturo Ardila Gomez, and Klas Sander. Other comments and contributions were received from Om Bhandari, Shenhua Wang, Geert Wijnen, Laurent Durix, and Eun Joo Allison Yi. Team assistance was provided by Rachel Susan Palmer, Michelle Lisa Chen, Comfort Olatunji, and Tara Nidhi Bhattarai. Anna van der Heijden provided technical editing for the report.

Support was also provided by various agencies in Bhutan, and the authors thankfully acknowledge numerous opportunities to meet with, interview, or otherwise receive information from Bhutan's Gross National Happiness

Commission, Ministry of Information and Communications, Ministry of Finance, Ministry of Economic Affairs, National Environment Commission, Thimphu Thromde, Bhutan Power Corporation, Bhutan Postal Corporation, Bhutan Bus Corporation, and the National Statistics Office of Bhutan.

This study and resulting report were supported by the Government of Australia's AusAid Trust Fund and the Korea Green Growth Trust Fund. The support from these funds is gratefully acknowledged.

Executive Summary

This report summarizes the results of a technical analysis of the Bhutan Electric Vehicle Initiative. The analysis, assuming three scenarios for the uptake of electric vehicles (EVs) in the country, introduces international experiences and lessons learned in EV programs and looks at required investments—for example, for EV charging facilities—as well as expected economic impacts and policy outcomes. Overall, while the EV initiative presents an interesting opportunity for providing green mobility in Bhutan—with possible outcomes including reduced fuel dependency and greenhouse gas (GHG) emissions—the implementation of any such EV program requires careful planning, awareness raising, policy incentives, and investments to develop the market, while building on global technology developments, government fiscal capacity, and travelers' behavior.

EVs Offer an Interesting Opportunity for Introducing Green Mobility in Bhutan

Bhutan is rich in hydropower resources, with activities in the hydropower sector contributing 12.5 percent of GDP and revenue from hydropower exports contributing 3.7 percent of GDP in 2013. These shares are expected to grow substantially as more hydropower projects are completed and start generating electricity. In 2012, about Nu 10,269 million (US$171 million)[1] in power was exported to India. At the same time, Bhutan has to import about Nu 6,331 million (US$106 million) of fuel every year, of which almost 90 percent is used for transportation. Reducing dependency on fossil fuel is one of the main goals of Bhutan's 11th Five-Year Plan, and using hydro-based electricity in the transportation sector is considered one of the policy options to achieve this goal.

The global market for EVs is still nascent but shows positive signs of development in the near future. The automotive industry has adopted electrification as a pillar of future drive train technology with an average penetration rate in 2012 of 0.6 percent of yearly new vehicle sales in the 10 largest markets and EV uptake is expected to continue to increase significantly. This penetration rate is generally indicative of a pilot stage technology prior to widespread adoption once costs decrease and availability improves. The International Energy Agency (IEA) has projected that by 2020 annual global EV sales will reach almost 6 million vehicles, or 5 percent of total passenger car sales.

In addition to research and development (R&D), public spending on EVs in EV leading countries is used for fiscal incentives and the development of EV charging infrastructure. In the top 15 countries with EV initiatives, most public funding is going to long-term R&D (over US$8 billion or Nu 0.13 billion), followed by fiscal incentives (US$3 billion or Nu 0.05 billion) and charging infrastructure (reaching US$1 billion or Nu 0.02 billion). In terms of investments in R&D and technology, Bhutan might best focus on learning from international experience and evolving in line with sector developments worldwide.

Several trends in the global EV market may accelerate EV uptake over the medium term. The first global trend is the ongoing improvement of battery technology, which is expected to increase the range of EVs. Current EV technology is best suited for customers driving less than 120 kilometers per day; only one manufacturer, Tesla, is currently producing vehicles with an electric range of about 400 kilometers—comparable to the range of conventional vehicles. International experience, however, has indicated that range is a critical factor in consumers' purchase decisions. The second global trend is the falling price of EV batteries. Over the last five years, battery prices, which form a large part of the cost of EVs, have dropped significantly. Finally, the third global trend is the expanding number of EV models. Although Nissan did not introduce the first modern commercially available EV—the Nissan Leaf—until 2010, by 2014 there were already 20 commercial EVs on the market. Many other car manufacturers also have an extensive R&D roadmap to expand EVs over the complete model range.

When the local market is developed at a pace consistent with global market trends, an EV program can benefit from ongoing improvements in EV technology and resulting cost decreases in the global market. Technological improvements in terms of better and less expensive batteries, additional range, new product offerings, and increased availability of affordable EV models will increase the attractiveness and financial competitiveness of EVs. These improvements in the global market will allow for an increase in EV uptake in the medium term with less financial support from the government.

The Context for Developing an EV Program in Emerging Markets Differs from That in Leading EV Countries

As with other green technologies, the high cost of EVs compared to conventional vehicles is one of the key challenges for successful uptake. The high cost of introducing a new green technology to achieve carbon, environmental, and health benefits is a common trade-off faced by all countries. For EVs, the World Bank (2011) estimated that the high cost of EV batteries can make purchasing EVs about 1.5–2.0 times as expensive as buying regular internal combustion engine (ICE) vehicles, while the total cost of ownership (TCO) (according to 2015 data for China) of EVs is also 15 percent above the cost of owning ICE vehicles. Only by the early 2020s is the cost per mile of EVs expected to be comparable to that of ICE vehicles. The initial purchase price is projected to be comparable with ICE vehicles only by the latter half of the next decade, or from

2025 onward, given that a 15–20 percent price reduction is achieved for EVs while gasoline vehicle prices remain relatively constant. How costs and benefits should be weighed, what the fiscal capacity of the government is to undertake an EV program, and how the EV agenda fits into a country's development priorities are common public policy issues that are debated in each country's particular context.

Different sectoral characteristics, for example in vehicle markets or provision of public transport services, set a different context for EV development in emerging EV markets compared to leading ones. In leading EV markets, the vehicle market is mature and vehicle ownership has saturated. In those markets, people buy EVs as a second car to serve shorter trips. In Norway, for example, 42 percent of households own more than one car. The transport sector is also well advanced, and key sectoral challenges are more related to environmental issues and achieving low-emission transport. In contrast, in developing economies like Bhutan, the development of the transport sector is at a different stage. In Bhutan specifically, because of the high cost of vehicles in general (due to high vehicle tax rates), a large part of the population cannot afford to buy a car. As the public urban transport system also has not yet been well developed, urban mobility in the country is still a primary issue. Providing access to safe, reliable, efficient, and affordable transport services for a large part of Bhutan's urban population remains a key challenge.

Fiscal and technical capacities are also key differentiating factors affecting the context for EV development in emerging economies. International experience suggests that large public support is needed to promote EVs, in particular through the provision of fiscal incentives and support for charging infrastructure investments and operations in the early adoption phase. In Norway, for example, the subsidy to an EV owner is an estimated US$20,000 (Nu 1,200,000) at the time of purchase and between US$1,000 (Nu 60,000) and US$3,700 (Nu 22,200) annually. High-income EV countries have more fiscal space and thus can afford to pay for subsidies, public investments, and building of the technical capacity of stakeholders involved in the introduction of the EV technology. In emerging economies, governments may have a more limited fiscal capacity (as a result of a more limited revenue base, scarcer public funds, and more urgent priorities), as well as more limited technical capacity to introduce new technologies.

Because EV Technology Is Still in an "Innovator" Stage, Bhutan Will Benefit from a Phased Approach

A target for EV uptake should take into consideration international replacement rates as well as local factors. One of the key aspects of designing an EV program is setting an EV uptake target that will help guide further EV program planning, design, and implementation, as well as convey the level of policy commitment and expected pacing of EV market development to market players and stakeholders. The uptake target will be the basis for deciding on other key elements of the EV program, such as fiscal incentives, plans for charging infrastructure, and grid and urban planning. Trends in international replacement rates, outcomes of a technical and economic assessment of the potential for EV uptake in the local context, and financial and technical capacity of the government and stakeholders to implement a program are all key factors to consider when setting the EV uptake target.

The uptake in the 10 largest EV markets is approximately 0.61 percent of new vehicle sales. Norway and the Netherlands saw the highest market penetration rates with replacement rates of 2.48 percent and 1.57 percent, respectively. Although the shift from conventional vehicles to EVs varies a lot between countries, key factors that have been observed to induce switching in EV leading countries are the prevalence of multicar households, a relatively high GDP per capita, a strong subsidy scheme, and the availability of a package of government incentives—both financial and nonfinancial (such as preferred parking and rights to use restricted lanes). Overall, it is expected that the uptake and potential market of EV buyers in Bhutan is smaller than those in advanced economies because of the country's different sociodemographic characteristics, such as relatively lower income, a lower percentage of people who own a family car, and overall a lower level of highly educated persons.

Adoption of EV is influenced by several key factors, such as the vehicle market, consumer preferences, and the availability of charging infrastructure. In addition to the factors that contribute to a shift from conventional vehicles to EVs (multicar households, high GDP, subsidy schemes), other key influencing factors include the availability of suitable vehicles, an attractive TCO, the availability of charging infrastructure, EV awareness, and suitability of user characteristics for switching to EVs. To assess the potential for uptake, a first factor to look at is whether travel demand in the Bhutanese context can be served by EV driving characteristics. Some key advantages of the Thimphu area for EV uptake include its relatively small geographical size and mild climate. These factors accommodate the trip types that largely align with the current offering of EVs. When switching to EVs, however, drivers may still need to adjust their behavior to accommodate for charging times, long distance trips (for example, to Paro and back), and available vehicle models. A study of EV use in the United States, for example, found that EV owners primarily use their EVs for short distance daily commutes, while other long distance trips might still require different vehicles or other modes of transport. A key difference, however, between the United States and Bhutan is that total car ownership in Bhutan is still low: in Bhutan less than 20 percent of households own a family car. Vehicle ownership in Bhutan is 70 motor vehicles per 1,000 people, compared to more than 500 vehicles (for example, 591 in Norway) per 1,000 people in high-income countries.

Successful EV programs focus on target groups, as different segments of a country's vehicle market may not be similarly suited for a switch to EVs. For example, not all vehicles with an ICE may be replaced 1:1 with an EV because of product uncertainties related to driving range, charging time, availability of charging infrastructure, and availability of suitable EV models. Although international developments will increasingly address these various limitations,

EV programs might specifically address suitable target groups for a transition from ICE to EV. In Bhutan, the government fleet, taxi drivers, and private car owners have been identified as potential target groups for a first phase of EV adoption.

In this report, three scenarios—covering low, high, and super high EV uptake—are used to assess the required levels of fiscal incentives, charging infrastructure needs, and policy impacts, which can help inform the government when setting an EV uptake target. In the low EV uptake scenario, with a replacement rate similar to the average international replacement rate of about 1 percent, by 2020 the number of EVs in Bhutan is expected to be about 500. In the high uptake scenario, with a replacement rate above that of leading EV countries, the number of EVs is expected to increase to about 1,500, while in the super high uptake scenario, with a replacement rate that matches the Royal Government of Bhutan's (RGoB) ambition to introduce 1,000 EVs per year, the number of EVs is expected to be about 6,000 in 2020.

Internationally, EV pilot programs in metropolitan areas are used to test the EV adoption approach before a countrywide implementation. Metropolitan areas are typically characterized by higher incomes, shorter travel times, and more incentives to improve air quality or reduce CO_2 emissions, making them more suitable for EV programs. Thimphu has these characteristics, which, along with its mild climate, make it a good pilot area for testing the new technology and building a foundation for a cost-effective countrywide rollout. EV adoption involves new stakeholders and know-how, and a pilot is key for testing how different aspects of the implemented EV solution perform, such as, for example, the operation of charging infrastructure (that is, solutions for apartment buildings and safety), efforts to change customer behavior, incentives for different target groups, and EV awareness campaigns.

Governments Use Fiscal Incentives to Enhance the Financial Attractiveness of EVs

An analysis of the TCO of EVs and ICE vehicles in Bhutan suggests that the TCO for ICE vehicles is more favorable than the TCO for EVs, even with the current incentives.[2] The key influencing factors determining the TCO for vehicles include (a) annual mileage (an average vehicle in Bhutan drives about 10,000 kilometers per year, while EV savings only start at 15,000 kilometers); (b) upfront costs (the upfront costs of an EV model are much higher—140–170 percent—than those of an ICE vehicle); (c) interest rates on vehicle loans (which are relatively high in Bhutan—about 14 percent—and thus add considerably to the overall financial costs given the higher upfront cost of EVs), and (d) fuel price (future fuel price increases will influence the TCO calculation and financial attractiveness of EVs). With the current incentives in place, the TCO of ICE vehicles is still more attractive than that of EVs, mainly because of the relatively moderate annual mileage of an average vehicle in Bhutan and the high upfront cost of EVs. Moreover, lower household income levels in the country, compared to high EV uptake countries, suggest that affordability will be an issue.

For taxis, switching to EV is more financially viable, but also more technically challenging. With the current incentive program, a strong financial case exists for the taxi market to switch vehicles with an average annual mileage of 50,000 kilometers to EVs. However, factors other than the TCO, some of which are unique to taxi operations, will also play a role. These include a taxi driver's ability to pay a higher upfront cost, access to vehicle financing, the need for fast charging during the day, parking availability in public housing areas, revenue impact, and marketing opportunities.

International experience suggests that the effectiveness of fiscal incentives varies as purchase decisions depend on both price and nonprice measures. Although financial incentives are essential for encouraging EV uptake, their effectiveness varies as consumer preferences differ; and nonprice factors, such as the convenience of charging infrastructure, also influence decision making. To address this, Norway, for example, augments its strong financial incentives with special lane access and parking perks, thus also providing benefits in terms of convenience and status. Internationally, evidence also exists that consumers value the perception of benefiting the environment when switching to EVs. In Bhutan, purchase decisions will thus also be determined by factors such as product risk perceptions, consumer preferences for new products, and environmental consciousness.

Globally, Charging Infrastructure for EVs Will at First Rely Mainly on Home and Workplace Charging Because the Business Case for Public Charging Requires More Public Support

The availability of charging stations is essential for EV uptake. In several countries, the availability of charging stations has been shown to have a direct and positive correlation with EV uptake. Worldwide, most charging infrastructure is and will continue to be at home and work, with fast charging available for occasional long-distance trips for passenger vehicles. Fast charging, however, is essential for taxis. The most cost-effective solution for charging infrastructure is private charging stations. For apartment buildings, dedicated parking with private chargers needs to be in place. A less economical solution is the use of public slow and fast charging points, as public charging requires the involvement of multiple stakeholders and more expensive charging stations with higher operational costs. The existence of public charging stations, however, will contribute to increased EV awareness.

International experience illustrates that creating a positive business case for public charging is difficult; public charging still requires some form of public investment or subsidy. Various approaches have been used to finance and build public charging infrastructure, with approaches depending on demographic characteristics, the road network, public investment, and strategic choices of regional and national governments. In the Netherlands, with one of the densest and best-used public slow charging networks in the world, trends in charging can be seen after five years of placing public charging points, but still no business case exists for private parties to become involved. Both the public and the private sectors are working hard to create this business case by developing cheaper techniques, using new pricing mechanisms, adjusting legislation, implementing tax reforms, and creating an innovative business model. With this combined effort, a business case for public charging is expected to become feasible in the coming years.

In developing a network of charging infrastructure, Bhutan would benefit from a phased approach. To reduce costs and limit the number of stakeholders involved, initial efforts might best focus on stimulating home and workplace charging. Fast chargers add additional value to the normal charging network. Moreover, a phased rollout, in pace with global EV developments and involving dedicated user groups, is necessary to gain experience and build a possible business case. For example, in the first five years, Bhutan could concentrate on creating a charging network only in densely populated areas with a current demand from EV drivers. For the fast charging network, a pilot area can be chosen that provides EV customers with direct benefits while the charging operator can derive some income. Building on those first experiences, additional public fast charging can be planned for more rural areas, once prices have fallen and more experience is available with different types of possible profitable business models through a combined effort of both the public and the private sectors.

Investment Needs for Charging Infrastructure and Potential Grid Impact Will Depend on the EV Uptake Target

The estimated infrastructure costs for the three scenarios range from US$0.68 million (Nu 40.8 million) in the low uptake scenario to US$11 million (Nu 660 million) in the super high uptake scenario, accounting for, respectively, 0.1 percent and 1.7 percent of the total national budget (Nu 38,843 million in 2011/2012). In the low uptake scenario, by 2020 about 648 normal chargers (home, work, and public) and about 10 fast chargers would need to be in place. In the super high uptake scenario, these numbers increase to about 7,000 normal and 240 fast chargers by 2020.

The impact of EV charging on the electricity grid will depend on the pacing of EV uptake; additional analysis and measures for off-peak charging may be needed. The level of EV uptake will impact the amount of electricity required from the grid. While in the low uptake scenario only 0.4 percent of the total peak demand will stem from EV charging, in the super high uptake scenario this increases to 5.2 percent. In particular over the medium and long term, the electricity grid must be ready to handle the impacts of EV uptake and adapt to accommodate the additional load. Off-peak charging will be essential for accommodating this increase in demand without having to overbuild infrastructure. Although a further detailed analysis is needed on the distribution level to determine the expected impact on the grid and judge the required investment, the present indication is that any small-scale uptake (that is, the low uptake scenario) is manageable with the grid "as is." However, the impact of a wider uptake (that is, the super high uptake scenario) will inevitably be noticeable, and measures for off-peak charging will need to be in place.

The Bhutan Electric Vehicle Initiative • http://dx.doi.org/10.1596/978-1-4648-0741-1

Effective Operation of Charging Infrastructure Requires a Dedicated Operator and a Market Model That Fits the Local Situation

A hybrid market model, involving both public and private parties, can allow for effective operations by a dedicated operator of the charging network infrastructure. A capable operator of the charging infrastructure is essential for safe, user-friendly, and reliable charging operations; the operator will need to plan and operate the home, public, and fast charging networks. Bhutan Power Corporation (BPC), which has relevant experience in electricity grid planning, has indicated the role of charging operator is outside its existing mandate and will for the moment not take the role. However, as public charging (unlike home and workplace charging) still requires public investments, a model that involves both public and private parties to cooperate and invest in charging infrastructure is preferred.

For charging facilities to operate effectively, sound contracting and financing agreements are essential. The contracting mechanism for the dedicated operator, agreed between the operator and the government, will need to cover, among other things, capital investments and asset management of the charging facilities, operation cost and cost recovery, the viability gap, and operation standards. In addition, building the institutional capacity of the operator and relevant stakeholders will be key, as the operation of charging facilities is new to Bhutan. The operator and stakeholders will need to work together closely, with, for example, the Gross National Happiness Commission (GNHC) addressing service pricing, the Ministry of Information and Communications (MOIC) contributing transport planning, BPC handling grid connection and electricity supply, and Thimphu Municipality managing land and planning. Key aspects of a successful operation of the charging infrastructure will involve obtaining and sharing the necessary technical expertise, a government collaboration approach, and—if an existing organization will be the operator—a possible structural adjustment of the organization to allow it to operate effectively in its new role.

More Ambitious EV Uptake Targets Will Require More Public Support and Higher Investments

To achieve the EV targets of each of the three uptake scenarios, different levels of public support and other measures are needed. As shown in table ES.1, public support is required for any EV program, mainly to support fiscal incentives and subsidize the investment and operation of charging infrastructure. However, the larger the intended level of uptake, the more public support and other measures are required. For public charging, depending on the nature of the contractual arrangement with a private operator, subsidies may involve an upfront subsidy as well as operating subsidies that are recurring over the years of operation. Any government funding strategy should carefully consider recurring costs in addition to the upfront capital subsidies.

Table ES.1 Summary of Key Assumptions of Each EV Scenario and Estimates of Public Support Required during 2015–2020

	Scenario 1—low EV uptake	Scenario 2—high EV uptake	Scenario 3—super high EV uptake
Key assumptions on uptake scenarios	Assume the EV replacement rate for private fleet and government is about 1% and for taxis 2%. Top 10 EV uptake countries have an average penetration rate of 0.6%.	Assume the EV replacement rate for private fleet and government is about 3% and for taxis 10%. Norway, the leading EV country, has a penetration rate of 2.5%.	The EV replacement rate for government and private fleet is 5% and for taxis 100%. It is the RGoB's ambition to introduce about 1,000 EVs per year.
Target number of EVs	79 EVs per year; 476 EVs in 2020	245 EVs per year; 1,472 EVs in 2020	1,022 EVs per year; 6,132 EVs in 2020
Key assumptions on potential for switching	The first group of consumers is likely to be less financially motivated. They will buy EVs because of EV product features, even though the TCO of EVs is higher than that of ICE vehicles. This is observed in many EV countries with an EV penetration rate around the average rate at less than 1%. In these countries, TCO of EVs is observed to be higher than TCO of ICE vehicles. EV for taxis it will be financially viable but technically challenging, and a basic charging network is required.	To reach more consumers, EV will have to reach more financially motivated consumers. This group will buy EVs when the TCO of EVs becomes lower than TCO of ICE vehicles. This is in line with the TCO observed in a leading EV country, i.e., Norway. EV for taxis will be financially viable but technically challenging, and a moderate charging network is required.	To reach even more consumers, EV will have to reach a group that is highly financially motivated. As this is the extreme case and international experience is not available to draw upon, it is generally assumed that this group will buy EVs if the sales price of EVs is below that of ICE vehicles and TCO of EVs is significantly lower than TCO of ICE vehicles. EV for taxis will be financially viable but technically challenging; an extensive normal and fast charging network is required with large geographical coverage.
Estimated fiscal incentives required to achieve target	Current incentives for private vehicles are likely to be adequate to meet the target.	It is roughly estimated that about 10% further price reduction (e.g., in the form of a cost subsidy) will make TCO of EVs the same as or lower than TCO of ICE vehicles.	It is roughly estimated that about 35% further price reduction (e.g., in the form of cost subsidy) will make sales price of EVs lower than sales price of ICE vehicles, which will in effect cause TCO of EVs to be significantly lower than TCO of ICE vehicles.
Total Fiscal Cost of Incentives Program (2015–2020) in million Nu (US$)	593 (10)	1,967 (33)	7,140 (119)
Total investment in public infrastructure (2015–2020) in million Nu (US$)	62 (1)	320 (5)	752 (13)
Other measures	Add some charging stations	Raise consumer awareness, develop the charging infrastructure network (fast/slow), and ensure more EV models	Develop exceptional consumer awareness, create a super infrastructure network (fast/slow/public), and ensure more EV models
Locations	Mainly Thimphu	Thimphu, and other major city centers	Countrywide
Target groups	Not specific	Government, taxi, private	All target groups; high percentage of taxi vehicles

Source: World Bank analysis.
Note: EV = electric vehicle; ICE = internal combustion engine; RGoB = Royal Government of Bhutan; TCO = total cost of ownership.

In providing fiscal incentives, fiscal support can be either direct (for example as cost subsidies) or indirect (tax exemptions). If fiscal incentives are limited to tax exemptions, fiscal support will be indirect in the form of uncollected tax revenue on EVs. Direct fiscal support, however, will require additional public funding. To estimate how much incentive is needed to meet the EV uptake targets of the three scenarios, the analysis assumed that the government will use cost subsidies in addition to the currently provided tax incentives as additional fiscal incentives to influence the price of EVs. Although it is difficult without additional information on consumer preferences to estimate what price reduction is needed to influence consumers to switch to EVs, for the purpose of the analysis it is roughly assumed that switching will occur if the TCO of EVs is below the TCO of ICE vehicles (see table ES.1). The larger the price gap between the two, the higher the potential for switching.

Based on the analysis, most of the public support for the EV program will be used to fund the fiscal incentives program to influence consumer purchases. Naturally, the higher the EV uptake target, the more public support is needed to finance the fiscal incentives. Table ES.1 illustrates that, based on current technologies, achieving an uptake rate of 79 vehicles per year in the next six years (2015–2020) may not require additional support because the market would capture buyers who are less sensitive to the price of the EVs. The needed fiscal support for this scenario during 2015–2020 would then include a tax waiver on EVs that is estimated to be about Nu 593 million (or about US$9.88 million) and an investment in infrastructure of about Nu 62 million (US$1.03 million). However, to achieve the goal of the high uptake scenario, the more aggressive uptake rate of 245 vehicles per year will require additional subsidies on top of the current tax exemption incentive, resulting in a total fiscal support of approximately Nu 1,967 million (US$32.78 million) and Nu 320 million (US$5.33 million) for infrastructure investment. Similarly, in the super high case scenario, with an introduction of 1,022 EVs per year, total fiscal support of Nu 7,140 million (US$119 million) over the six-year period for the incentive program and Nu 752 million (US$12.53 million) for infrastructure investment will be required. If the upfront costs of EVs will be around 35 percent lower, these figures are reduced by roughly Nu 149 million (US$2.48 million) and Nu 871 million (US$14.52 million) for the high and super high uptake scenarios, respectively. For the low uptake scenario, fiscal support would remain the same because for this scenario an additional cost subsidy is not required.

The fiscal impact of the EV program will depend on the ET uptake target and level of financial incentives. In the low EV uptake scenario, the annual uncollected tax revenue from the current incentives (the three tax exemptions) is likely in the range of 0.3–0.45 percent of annual tax revenue. However, the fiscal impact in the high and super high scenarios will be more significant, ranging from 0.99 to 5.47 percent of annual tax revenue for 2015–2020. It is important to note that this current tax regime may have the effect of encouraging more widespread purchase of conventional vehicles from India because these are not subject to customs duty (which at 45 percent is the same as the sales tax). It is difficult to project what impact Bhutan's new vehicle tax regime will have on the overall demand for vehicles and what fiscal consequences this may have.

Government fiscal limitations may imply that international grants and user charges will be necessary. Although the government's fiscal deficit is projected to remain low, Bhutan relies heavily on external grants and loans to finance government expenditures (World Bank 2014). Domestic revenues now cover about 67 percent of total expenditures; although 52 percent of current expenditures and 31 percent of capital expenditures are funded by domestic resources, the rest is funded by external grants and loans. This implies that the government may have fiscal limitations to implement a new initiative or capital investment by relying on general tax revenues. External grant resources will likely be a major funding resource for the EV program. At the same time, user charges for beneficiaries should be considered to fill in the financing gap.

A phased approach to infrastructure investments, financial incentives, and nonprice incentives will increase the chance to identify the right mix and level of incentives to achieve EV policy targets. Given that Bhutan's EV market differs from markets in developed economies, careful consideration must be given to its specific implementation challenges. Most likely, these considerations will lead to an adjustment of the initial mix of price and nonprice incentives that the country can offer. In particular the deployment of charging infrastructure should reflect a demand-driven approach, and efforts to switch over the taxi fleet should also be carefully considered because this—as international experience suggests—can be particularly technologically challenging. In the case of charging infrastructure, Bhutan's first choice might be to go for predominantly private charging (that is, at home or work). For high mileage users like taxis, dedicated fast charging facilities will need to be in place, with the number of facilities in line with the number of EVs. The number of public fast charging stations for range extension, mostly along the highway to accommodate longer drives, can be kept to a minimum in the low uptake scenario.

Public Awareness and Stakeholder Involvement Are Important to Kick-Start the EV Program

Increased public awareness and visibility combined with objectivity will help communicate accurate information to potential buyers and the general public. EVs are new to many potential customers in Bhutan. As with new technologies in general, increased public awareness is needed to overcome various biases or reservations, such as anxieties or misinformation about the EV driving range, battery life, EV battery recycling, and the convenience or inconvenience of charging. Any effective EV development strategy will need a robust communication component. One important caveat, however, is the potential for reputational risk if consumers perceive that EV-related information is not entirely accurate. A communication campaign that focuses on giving potential consumers objective facts to inform their purchasing decisions rather than aggressive marketing of EVs is highly advisable.

The Bhutan Electric Vehicle Initiative • http://dx.doi.org/10.1596/978-1-4648-0741-1

Overcoming the financial, economic, and political barriers in setting up an EV program will require a combined effort from all involved stakeholders. Despite the growth of Bhutan's EV market, overall awareness and knowledge of the various stakeholders is still very limited. International experience shows that for a successful EV implementation, stakeholders from both the public and the private sectors need to work together (as public-public, private-private, or public private partnerships [PPP]) on different levels to move the EV sector to the next phase.

International and domestic long-term knowledge exchanges are essential for effective uptake. The field of electric mobility is relatively new and touches on various policy areas, including sustainability, mobility, urban planning, and energy. Best practice experiences from other countries can greatly assist policy makers in Bhutan. Moreover, the quick developments in the EV market will necessitate ongoing assessments and possible adjustments of relevant policies.

To Achieve the Policy Impact Aggressive (and Costly) Scenarios of EV Adoption Need to Be Aimed For

Key government policy objectives can be achieved only under the super high EV uptake scenario, but such a level of ambition will be costly, technically challenging, and subject to the consumer response to EVs. The persistent current account deficit with India is a major macroeconomic issue for Bhutan, and the government is facing a critical challenge of curtailing imports and substituting imports with domestically available resources where possible. In addition, reducing fuel dependency and increasing environmental benefits such as reduced carbon emissions from fossil fuel consumption are also important broader objectives of the EV policy. The analyses of the policy impact of the three scenarios (table ES.2) suggest that only the super high uptake scenario will be able to achieve significant policy impact in terms of fuel import reduction, avoided GHG emissions, and improved trade balance. However, realizing this scenario will have strong fiscal implications because fiscal support for this scenario is estimated to reach 3.6–5.5 percent of annual taxes revenue. The savings and benefits will also not impact the same institutional groups. The financial savings will be accrued to households who enjoy EVs at a subsidized price, as well as savings in their fuel expenses. Thus, the savings also imply an implicit transfer from the public to households. The country will enjoy the benefits of avoided fuel imports and the social value of avoided GHG emissions at the cost of providing fiscal incentives to EV buyers.

The reduction of gasoline imports could be significant in the high uptake scenario, but the overall impact on fuel imports will not be as significant because diesel comprises the larger share of total fuel imports. Overall, 70 percent of Bhutan's fuel imports involve diesel, which is used primarily for freight transport vehicles. The current global technological offering for EV does not overlap with this class of conventional vehicles. In the high uptake scenario, the annual avoided fuel imports in 2027 will be about 5.1 percent of total fuel imports in

Table ES.2 Summary of Policy and Fiscal Impact in Three Scenarios

	Scenario 1—low uptake, million Nu (US$)	Scenario 2—high uptake, million Nu (US$)	Scenario 3—super high uptake, million Nu (US$)
Aggregated impact			
Aggregated benefits	799 (13)	2,580 (43)	19,282 (321)
Accumulated avoided fuel imports (2015–2027)	768 (13)	2,479 (41)	18,531 (309)
Accumulated avoided GHG emissions (2015–2027)	31 (1)	100 (2)	750 (13)
Magnitude of impact			
Annual avoided fuel imports in 2027 (% of total fuel imports in 2012)	1.57	5.1	38
Net impact on imports[a] during 2015–2027 (Max. annual import reduction as % of annual imports)	0.07	0.24	1.81
Annual GHG savings (as % of annual transport emission in 2000)	1.39	4.5	33.6
Annual fiscal impact to government during 2015–2020 (as % of annual taxes revenue)	0.30–0.45	0.99–1.51	3.60–5.47

Source: World Bank analysis.
Note: The study assumes that the EV initiative will be implemented in six years from 2015 to 2020, which means the fiscal impact will incur during this 2015–2020 implementation period. It is further assumed that EVs will be used for eight years, which means the benefits from EV will incur during a 2015–2027 impact period. EV = electric vehicle; GHG = greenhouse gas; ICE = internal combustion engine.
a. The net impact on imports is calculated based on the reduction in fuel imports, the increase in imports for charging equipment, and the increase in imports as a result of the higher price of EVs compared to ICE vehicles.

2012, or 20 percent of gasoline imports in 2012. In the low uptake scenario, the annual avoided fuel imports in 2027 will be about 1.6 percent of total fuel imports in 2012, or 6.1 percent of gasoline imports in 2012. In the super high uptake scenario, the annual avoided fuel imports in 2027 will be about 38 percent of total fuel imports in 2012, or 147 percent of gasoline imports in 2012. Total avoided fuel imports during 2015–2027 are estimated to be in the range of Nu 768–18,531 million (about US$12.8 million–US$308.85 million) between the low and super high scenarios. However, achieving the higher levels of reduction (as under the high and super high scenarios) would, in particular, require high mileage vehicles such as taxis to switch from ICE to EV. In addition, the estimate of avoided fuel imports is highly dependent on changes in the global oil price.

The potential of the EV policy to contribute to reducing the total trade deficit is more limited. The net incremental trade impact will be mixed because the reduction in imports as a result of fuel savings is offset by incremental increases in imports due to the higher initial investment required for EVs compared to that for ICE vehicles. The contribution of the EV policy to reducing total imports is also lessened considering that EVs will have to be replaced periodically at the end of the vehicle life, resulting in recurring incremental increases in imports at every replacement period. The impact will also be highly sensitive to fuel price increases in the future as well as the cost of EVs. The net import reduction impact can be sustained only in the super high uptake scenario. Nevertheless, the magnitude of any impact is likely to be small (less than 2 percent) given the potential size of the EV uptake targets compared to the overall imports value.

Annual GHG reduction benefits in 2027 are estimated to be 1.39 percent, 4.5 percent, and 33.6 percent of annual transport emissions in 2000 in the low, high, and super high uptake scenarios, respectively. In terms of total accumulated savings during 2015–2027, the estimates for the three scenarios are 17,276 tons of carbon dioxide equivalent (tCO_2e), 55,774 tCO_2e, and 416,897 tCO_2e in the low, high, and super high uptake scenarios, respectively. In the high uptake scenario, the annual avoided GHG emission in 2027 will be 5,312 tCO_2e. At a social value of US$30 (Nu 1,800) per ton of carbon emissions in 2015, the total accumulated social value of avoided carbon emissions during 2015–2027 is estimated to be Nu 31 million (US$0.52 million), Nu 100 million (US$1.67 million), and Nu 750 million (US$12.5 million) in the low, high, and super high EV uptake scenarios, respectively.

Sound Selection and Regulation of Charging Technologies and Product Standards for EVs Are Important for Long-Term EV Market Development

Investment in charging technologies should align with those adopted in neighboring countries. Charging EVs with a normal household plug is unsafe, and unintended consequences of this approach can even damage public perceptions about EV. Specialized technologies for both household and public fast charging facilities are required, and Bhutan must seriously consider which technologies are best to adopt. Given the scale of Bhutan's market, the most likely solution will entail adopting existing foreign standards. Moreover, the government should consider the decisions of both China and India, whose automotive industries will likely play significant roles in EV development in the coming decades. India is currently still deciding on its EV technology standards; a decision is expected in 2016.

Basic minimum standards must be enforced to protect consumers from EV products that may not be safe or reliable. Not all EVs have the same quality across manufacturers. In particular, battery performance and battery service life can vary considerably. Some countries include targeting mechanisms for product standards in their incentive programs to ensure that public funds do not support lower quality vehicles.

Utilities and car manufacturers globally are piloting several promising solutions for battery second life, but no clear solution is yet in place. Batteries for EVs are improving considerably, with a current benchmark warranty of 100,000 miles and eight years. Even after the warranty has expired, the vehicle can be used for years, but with a limited and declining range. In the end, the battery might have to be replaced, with the old battery possibly reused for energy storage or as a backup power source in rural areas. Alternatively, the battery can be recycled (either certain battery modules or all materials in the battery).

EV Is One Measure for Achieving Inclusive Green Transport and Is Normally Pursued in Parallel with Other Priority Measures

EV will benefit richer households, and those households will contribute to the environment. Because of the high cost of EVs, EVs will be affordable only to higher-income households, which means any EV subsidy program will specifically benefit this higher-income group. The EV buyers, however, will also contribute to public benefits in terms of reduced fuel imports, improved urban air quality, and reduced GHG emissions. These public benefits should be balanced with the distributive impact of the indirect subsidy to EV buyers.

Public transport improvement could be pursued in parallel with an EV policy, as it will directly serve those lower-income users who have no other choice than to use public transport. Addressing public transport will result in a pro-lower-income distributive impact because public transport is used mainly by the poor and vulnerable groups without access to cars. Improved public transport that is designed to connect travelers from their homes to work and community services would be expected to result in a switch from car use to public transport by providing substantial welfare benefits (that is, savings in time and out-of-pocket expenses) along with significant "green" benefits; as such, improvements in public transport might well be considered an essential component of a comprehensive approach to developing green transport in Bhutan.

Notes

1. This report uses an exchange rate of 60 Nu (Bhutanese ngultrum) = 1 U.S. dollar (US$).
2. The current incentive program comprises exemptions of the sales tax and customs duty, which are normally imposed on vehicles. In addition, ICE cars will also be subject to the green tax, but EVs will not.

References

World Bank. 2011. *The China New Energy Vehicles Program: Challenges and Opportunities.* Washington, DC: World Bank.

———. 2014. *Bhutan—Development Update.* Washington, DC: World Bank.

Abbreviations

A	ampere
AC	alternating current
A/C	air-conditioning
BEA	Bhutan Electricity Authority
BEV	battery electric vehicle
BMS	battery management system
BPC	Bhutan Power Corporation
BSS	battery swap station
CO_2	carbon dioxide
CPI	consumer price index
DC	destination charging
DoD	depth of discharge
DSO	Distribution System Operator
EV	electric vehicle
EVI	Electric Vehicle Initiative
EVSE	electric vehicle supply equipment
FEV	full electric vehicle
FYP	five-year plan
GDP	gross domestic product
GHG	greenhouse gas
GNHC	Gross National Happiness Commission
GPS	Global Positioning System
GSURR	Social, Urban, Rural & Resilience Global Practice
GTIDR	Global Practice of Transport & Information and Communications Technology
GWh	gigawatt-hour
h	hour
HH	household
HOV	high-occupancy vehicle

HSD	high-speed diesel
ICE	internal combustion engine
IEA	International Energy Agency
kW	kilowatt
kWh	kilowatt-hour
LCA	life cycle analysis
Li-ion	lithium-ion
LUCF	land use change and forestry
MOIC	Ministry of Information and Communications
MW	megawatt
NEDC	New European Driving Cycle
NPR	Nepalese rupee
NSB	National Statistics Bureau
Nu	Bhutanese ngultrum
NUS	National Urban Strategy
OC	opportunity charging
OEM	original equipment manufacturer
PHEV	plug-in hybrid electric vehicles
PPP	public-private partnership
PRTM	management consulting subsidiary of PricewaterhouseCoopers
R&D	research and development
RFID	radio-frequency identification
RGoB	Royal Government of Bhutan
RSTA	Road Safety and Transport Authority
SOA	safe operating area
SOC	state of charge
t	ton (1000 kg or 1 metric ton)
TCO	total cost of ownership
tCO_2e	tons of carbon dioxide equivalent
US$	United States dollar
VAT	value added tax
WTW	well-to-wheel

CHAPTER 1

Introduction

The Bhutan Electric Vehicle Initiative in Context

The Royal Government of Bhutan (RGoB) has announced its Electric Vehicle (EV) initiative. The initiative is being developed in line with the country's commitment to address environmental issues and reduce dependency on fossil fuel, as described in Bhutan's 11th Five-Year Plan. This report, which has been prepared at the request of the RGoB, presents a technical analysis of key technical, financial, economic, and policy issues relevant for designing and implementing an EV initiative in Bhutan, highlighting in particular international experience with EVs and EV programs.

Globally, the use of EVs is growing steadily as technological advances improve performance and cost competitiveness, while governments adopt policies to promote EVs under a mix of objectives including energy security, reduction of greenhouse gas (GHG) emissions, climate change mitigation, local air pollution control, and domestic industry development. In recent years, most large original equipment manufacturers (OEMs) have announced or have already introduced EVs. Moreover, as prices of batteries and other components are falling and faster charging techniques are being introduced, significant growth rates for EVs have been seen in several European countries and China, Japan, and the United States. Key factors that have been shown to determine the uptake of EVs are household characteristics, existing travel behavior, availability of vehicle technology and charging infrastructure, financial incentives, knowledge on EVs, and customer perceptions. Also, during this early stage of EV uptake, it is clear that all EV markets require significant public support to make EVs a viable alternative to traditional ones.

Many countries have set clear goals for EV adoption. In October 2014, for example, the European Union published directives on the deployment of infrastructure for alternative fuels, setting out minimum requirements for its member states to build up infrastructure, including charging points. While some countries such as China, Germany, Japan, and the United States have clear strategic reasons to invest in EV research and development (R&D) to develop their domestic auto industries, other countries, most notably Norway, the Netherlands, and the United Kingdom, are pursuing EV programs and policies in light of ambitious climate and environmental targets and to move toward zero-emission transport. Local governments, such as those in Beijing, California, and London, are also supporting EVs to address local air pollution that is affecting the health and well-being of residents. Although in developing economies priorities differ from those in high-income countries, energy and environmental challenges remain important. In Bhutan, as in many countries, sustainable transport development is intertwined with broader development issues such as the trade deficit and macroeconomic management, energy security, and green growth.

Bhutan's unique context offers both inherent advantages and disadvantages to implementing an EV program, thus posing both challenges and opportunities. This report, prepared by the task team of the World Bank Technical Assistance Project on the Bhutan Green Transport and Electric Vehicle Initiative, is based on international experience with electric vehicles and EV programs and summarizes key success factors and lessons applicable in the context of Bhutan. The report, considering three different EV uptake scenarios, specifically analyzes infrastructure requirements, required levels of fiscal incentives, and possible economic impacts. The report uses available data and information from current literature, as well as information collected in interviews or during the task team's visits to Bhutan. The focus of the report is on passenger and light commercial vehicles.

Overall, the report aims to inform the RGoB and its decision-making process for EV policy development and implementation by summarizing relevant technical background information and presenting analytical work based on international experience, to provide a strong technical basis for deliberating the EV initiative and designing an optimal policy and implementation program from a technical, economic, and social point of view. In addition, this report also intends to contribute to knowledge sharing and policy discussions among relevant stakeholders.

Following this introduction, the next chapter (chapter 2) will present background information on the macroeconomic situation of Bhutan, as well as the national development strategy and sectorial context of the urban, energy, and transport sectors, to place the EV initiative in perspective. Next, chapter 3 provides an overview of the global situation with EV uptake, highlighting the key influencing factors and main target groups. The chapter will describe three possible scenarios for EV adoption in Bhutan, based on international experience and the aspirations for EV uptake by the RGoB. Chapter 4 analyzes global EV market and technology developments, as well as specific technical issues—such as batteries, effectiveness of the drive train, and drive range—in the context of Bhutan. Chapter 5 then reviews the available international experience with the use and effectiveness of fiscal and economic incentives. The chapter also analyzes the financial viability of EVs in the Bhutanese context and calculates the level of incentives needed to meet the targets in different EV scenarios. Chapter 6 focuses on planning for the charging infrastructure network; it assesses the needed infrastructure for the introduction of EVs and shows the different types, standards, locations, and stakeholders involved in the EV business. A detailed analysis is also provided for the specific situation of charging in Bhutan, looking at locations, expected numbers, market models, roles, financing, and the possible impact on the electric grid. Chapter 7 evaluates the broader fiscal and economic impacts of the EV initiative, including on trade, fuel dependency, and the environment. Finally, chapter 8 provides a quick assessment of stakeholders, distributive implications, and policy options to engage different stakeholders in the EV program, also describing how electric vehicles could contribute to a broader inclusive and green transport agenda.

CHAPTER 2

Background

Key Messages

- Bhutan's economy is on the rise and is driven largely by hydropower development, tourism, and services.
- Green growth is at the heart of Bhutan's development goals, and the national strategy calls for ecologically balanced sustainable development.
- Bhutan currently runs a large and growing account deficit due to increasing demand for imports, which poses a key macroeconomic management issue.
- Bhutan's fiscal position is sound with prudent management of spending, but tax collection could be improved to increase revenue.
- Improving the availability and quality of urban transport services in Bhutan is important. To accommodate the growing urban populations and increasing levels of motor vehicle ownership, the main urban centers of Bhutan (Thimphu and Phuentsholing) will need to focus on expanding the availability and quality of urban infrastructure and services, including transport services.

Bhutan's Macroeconomic Situation, Development Objectives, and Key Sectors

GDP Growth and Current Account Deficit

After a policy-engineered slowdown in 2012, which saw GDP growth decline to 4.8 percent (the lowest since 2008), in 2013 the economy in Bhutan managed to rebound to a GDP growth of 6.5 percent, supported by hydropower construction and higher electricity and food production following favorable rains. The macroeconomic projections for 2014 set GDP growth at 7.3 percent, which would stem from new projects, increased tourism receipts, easier credit conditions, and the effects of the Economic Stimulus Plan (World Bank 2014). One of Bhutan's key macroeconomic issues is its large and growing current account deficit, which was estimated at about 25 percent in 2013. Currently, the account deficit is projected to deteriorate over the medium term because of strong growth in import demand associated with the construction phase of the hydropower projects, as well as hydro debt service. When the hydropower projects are completed, however, electricity exports are likely to more than triple from current levels. The current account deficit should thus decline over the longer term, leading to a balance of payment surpluses starting in 2020.

Tight public spending has kept the fiscal deficit in Bhutan below 5 percent of GDP, despite a sharp decline in revenues. Overall, the fiscal deficit is estimated at 4.5 percent of GDP in 2013/14. The government continues to rely heavily on foreign grants to finance its expenditure; in FY2012/13, grants financed about 36 percent of total spending (10 percent of GDP), with 70 percent of grants coming from the government of India.

Eleventh Five-Year Plan: Self-Reliance and Green Socioeconomic Development

The Royal Government of Bhutan (RGoB) recognizes that Bhutan is facing both the challenges and opportunities of the 21st century in adapting social, environmental, and economic integration to a more globalized world. The recently announced 11th Five-Year Plan (FYP) for 2013–2018 (Gross National Happiness Commission, Royal Government of Bhutan 2013) focuses mainly on self-reliance and inclusive green socioeconomic development. Even with the robust growth rates of the economy over the past decade, Bhutan's dependence on imports, its narrow tax base, large dependence on hydropower revenues, and low levels of productive employment make self-reliance a major development target for the country. In addition to self-reliance, as a crosscutting principle the FYP also calls for the adoption of rigorous environmental standards and mainstreaming of green or carbon-neutral strategies in all activities, specifying the promotion of electric vehicles as a strategy to address environmental issues and reduce dependency on fossil fuel.

Energy Sector and Hydropower[1]

The energy sector is very important for the development of Bhutan and has been a primary focus in recent FYPs. In particular the development of hydropower has been driving economic growth and is currently a key source of income, accounting for about one-fifth of Bhutan's GDP and about 30 percent of total government revenues (International Monetary Fund 2014). Bhutan has an estimated hydropower potential of 23,760 megawatts (MW) with a mean annual energy production capability close to 100,000 gigawatt-hours (GWh). At present, the installed hydropower capacity is about 5 percent of the total potential (Gross National Happiness Commission, Royal Government of Bhutan 2013). Electricity tariffs are the lowest in the region, and a free electricity program for the first 100 units is being implemented from 2013 to 2015 to encourage rural communities to use electricity instead of firewood because of environmental and health concerns (Gross National Happiness Commission, Royal Government of Bhutan 2013).

Urban Development

Bhutan is undergoing a profound and rapid demographic transition from a largely subsistence, rural economy to an urban society. The 2008 National Urban

Strategy (NUS) assumed as the most likely scenario that by the year 2020 the nation will be 60 percent urbanized, adding an estimated 250,000 urban dwellers compared to the 30.8 percent urbanization level in 2005. The cities Thimphu and Phuentsholing are the two most popular destinations for urban migrants. Thimphu, the capital and main city, has already experienced a rapid increase in its urban population, which was followed by a rapid expansion of its urban extension; by 2040, the city is expected to again have doubled its population. Phuentsholing is an industrial hub linking Bhutan with its neighbor India. In anticipation of the growth of the urban population, levels and quality of infrastructure and services would need to be increased.

Urban Transport

In recent years and following the rapid urbanization rate, urban transport has undergone significant changes. Between 1990 and 2010, the number of Bhutanese living in urban areas nearly tripled, and rates of motor vehicle ownership have increased along with this trend. This has led to traffic congestion, particularly in the main urban areas such as Thimphu, as well as to other impacts on the urban environment such as increased emissions and reduced pedestrian safety. Bhutan's urban transport systems are still developing to meet the increasing demand for urban mobility. As of September 2014, total vehicle registration was 68,744 in Bhutan, with an estimated 38.4 percent of registrations in Thimphu Dzongkhag[2] based on the household ownership rate. During 2008 to 2012, the motorization rate grew about 12 percent per year, with an average of about 6,300 newly registered vehicles each year. Appendix A includes more detail about urban transport in Bhutan.

Global EV Initiatives and the Context of the Bhutan EV Initiative

EV technology has only recently been introduced to the global market, and the EV initiative of the RGoB will be one of the first in the emerging markets. Internationally, countries have initiated EV policies with a mix of objectives, such as to enhance energy security, reduce GHG emissions and local air pollution, and promote domestic car industries. In fact, EV initiatives in the five countries with the highest EV stock by the end of 2014 (China, Japan, the Netherlands, Norway, and the United States) are driven by clear strategic policy objectives or national mandates. Among this top five, three countries—China, Japan, and the United States—are global leaders in auto manufacturing with clear strategic reasons to invest in research and development (R&D) to maintain their industry's competitive advantage. In comparison, European countries are driven more by their climate and environment policies with committed targets for GHG emission reduction. The Norwegian government, for example, has adopted a target to reduce average carbon dioxide (CO_2) emissions for new passenger cars to 85 g/km by 2020 in order to meet a target of reducing GHG emissions in the Norwegian transport sector by 2.5–4 million tons. EV initiatives are also used by local governments such as those in Beijing, California, and London to address local air pollution. California's Zero Emission Vehicle Program chiefly aims to address local air pollution; the regulation requires major car manufacturers to increase the number of electric battery and fuel cell electric vehicles in the state.

In the case of Bhutan, the RGoB in its recent FYP announced its vision for Thimphu to become an EV city. The FYP explicitly stated that the "promotion of electric vehicles will be pursued to address environmental issues and reduce dependency on fossil fuel" (Gross National Happiness Commission, Royal Government of Bhutan 2013). As such, the policy environment for EVs in Bhutan is unique compared to that of other countries, as the EV program in Bhutan is pursued in the broader context of the country's external vulnerabilities (such as a fuel dependency risk), as well as to fulfill its environmentally sustainable development vision.

Notes

1. The information about hydropower is drawn from the Green Growth Report.
2. Thimphu Dzongkhag refers to the administrative district that includes Thimphu City.

References

Gross National Happiness Commission, Royal Government of Bhutan. 2013. *Eleventh Five Year Plan Volume I: Self Reliance and Inclusive Green Socio-Economic Development.* Thimphu, Bhutan: Gross National Happiness Commission, Royal Government of Bhutan.

International Monetary Fund. 2014. *2014 Article IV Consultation—Staff Report; Press Release; and Statement by the Executive Director for Bhutan.* International Monetary Fund Country Report 14/178. Washington, DC: International Monetary Fund.

World Bank. 2014. *Bhutan—Development Update.* Washington, DC: World Bank.

CHAPTER 3

Scenarios for Electric Vehicle Uptake in Bhutan

Key Messages

- Electric vehicle (EV) uptake in the 10 countries with the largest EV markets is about 0.61 percent of new vehicle sales. These countries often have a relatively high GDP per capita and a package of government incentives (financial and nonfinancial) in place.
- Influencing factors for EV adoption include the availability of suitable vehicles, an attractive total cost of ownership (TCO) for the EVs, the availability of charging infrastructure, suitable user characteristics, and raised EV awareness.
- The government fleet, taxi drivers, and private car owners are potential target groups for a first phase for EV adoption in Bhutan.
- Based on a theoretical calculation, in the low EV uptake scenario, about 400 EVs are expected to be adopted by 2020; in high EV uptake and super high EV uptake scenarios, these numbers would be about 1,500 and 6,000 electric vehicles, respectively.

Global EV Penetration

In most countries, electric vehicles are still a nascent and small segment of the overall market for vehicles. By the end of 2012, the global stock of electric vehicles was over 180,000, representing 0.02 percent of the total number of passenger cars. The International Energy Agency (IEA) has projected that this global stock in electric vehicles will reach 20 million vehicles in 2020, then accounting for 2 percent of all passenger cars.

In the 10 largest EV markets, the average rate of uptake is about 0.61 percent of new vehicle sales (figure 3.1). Only the Netherlands and Norway saw above-average market penetration rates of 2.48 percent and 1.57 percent, which might be explained by the prevalence in those countries of multicar households, along with relatively high per capita incomes, the presence of nonfinancial incentives (such as preferred parking and rights to use restricted lanes), and a strong

Figure 3.1 EV Sales Volume and Share of Total Vehicle Sales in Top Ten of Leading EV Markets, 2012

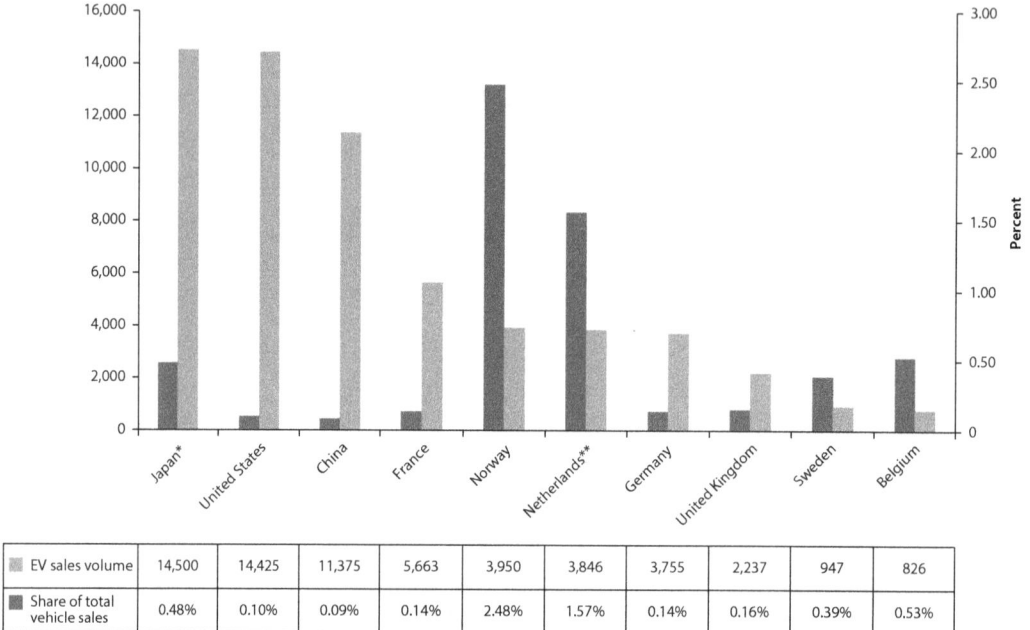

	Japan*	United States	China	France	Norway	Netherlands**	Germany	United Kingdom	Sweden	Belgium
EV sales volume	14,500	14,425	11,375	5,663	3,950	3,846	3,755	2,237	947	826
Share of total vehicle sales	0.48%	0.10%	0.09%	0.14%	2.48%	1.57%	0.14%	0.16%	0.39%	0.53%

Source: Hannisdahl, Malvik, and Wensaas 2013.
Note: Includes battery electric vehicles (M1, L7e). EV = electric vehicle; PHEV = Plug-in Hybrid Electric Vehicle.
* Best guesstimate.
** Also PHEV.

subsidy scheme. In California, the government uses a target of 15 percent EVs for all new sales in 2025, which means one electric vehicle is sold for each six regular ICE (internal combustion engine) vehicles sold; the state has reserved US$350 million (Nu 5.83 million) per year for six years to be used for incentives.

Influencing Factors for EV Adoption

The adoption of EVs varies a lot between countries and within segments of a country's vehicle market. Several factors that influence the rate at which users change from driving ICE vehicles to using electric vehicles include the availability of suitable cars, user characteristics, driving patterns, and access to charging points and parking. In addition, TCO, consumer awareness and trust, and visibility and marketing play key roles. This section discusses these factors in the context of Bhutan.

Availability of Suitable Vehicles

Currently only a few EVs are available in the global market. All major car manufacturers have at least one EV in production or planned, but available vehicles currently do not match all types of ICE vehicles in terms of available sizes, capacities, ground clearance, electric ranges, and price. Most full EVs are developed for use in urban areas, and currently no EV is available that offers the kind of ground clearance that would be required for driving on all of Bhutan's roads, which means EV adoption in Bhutan must focus on the urban areas. Vehicles in Bhutan also often carry heavy loads. Not all EV models offer sufficient trunk space to accommodate this.

User Characteristics

Globally, at the current stage of EV technology development, EVs are mainly bought by people with certain sociodemographic characteristics (Hjorthol 2013). As the Institute of Transport Economics demonstrated in a study published in 2013, "Early adopters of Electric Vehicles (EVs) are middle aged, between 30 and 50 years of age; the majority are men, they have high education and income, live in the vicinity of cities and belong to households with more than one car" (Hjorthol 2013).

Looking at the sociodemographic characteristics in Bhutan, about 50,000 people (or 9 percent of the population) are 30–50 years old and live in the urban areas (National Statistics Bureau and Asian Development Bank 2012). In addition, the mean annual household income in urban areas is Nu 282,671 (US$4,700); 6 percent of the total population has an education beyond grade 12; and the urban population is about 14 percent of the total population. Based on this information and the Norwegian study (Hjorthol 2013), some of the urban population in Bhutan with higher education levels and income could be a potential target group of "early adopters" (see also figure 4.2), as it is most likely they would be able to afford an EV. However, most people in Bhutan do not fall into any of these categories and would likely only switch to an EV when strong government incentives are in place. Although Bhutan has potential EV buyers, the number of actual buyers is expected to be very low compared to developed countries because of lower overall incomes and the fact that only 19 percent of households in Bhutan own a family car.

Driving Pattern

An EV driving pattern has to fit a user's existing driving pattern to allow a smooth transition. While a delivery company with services within a city area can easily accommodate the limited range of an EV, a taxi driver on long-distance routes will not be able to offer the same services when long charging times have to be taken into account. Within Thimphu, the maximum distance from north to south in the city is about 10 km, which is ideal for electric vehicles.

Access to Charging and Parking

Access to charging and parking is also very important for a smooth transition. For example, an employee with private parking at home and an office location with a company charger will be more confident to make the switch to EVs. Also, if a countrywide fast charging network is available, users will need to worry less about the driving range and can use their vehicles for longer distances (Anegawa 2010). Chapter 6 will specifically address parking and charging infrastructure.

Attractive TCO

The TCO for an EV compared to an ICE vehicle is a major factor in the purchase decision. The buyer needs to be able to get the right information, understand the calculation, and also trust the data when buying a vehicle—the EV—with a higher upfront cost (see chapter 5 for a calculation of TCO for Bhutan).

Consumer Awareness and Trust

In general, consumers will switch to EVs only when they trust the product.[1] At the moment, overall consumer awareness is low; and the general public and target groups such as taxi drivers are not well informed about the advantages or disadvantages of EVs. An independent information source that could provide fact-based information about EVs does not yet exist. A household survey might provide valuable information about the level of consumer awareness and what additional information may be required. Box 3.1 highlights the findings from a brief, first assessment[2] of EV awareness in Bhutan.

Visibility and Marketing Impacts

Internationally, corporate owners of vehicle fleets switch to EVs partly for marketing reasons, as the use of EVs shows an organization's corporate social responsibility. Because of this, visibility of EVs is important and a factor for switching to EVs. Visibility and marketing reasons, however, are not the only considerations for switching to EVs. According to a longitudinal study (California Center for Sustainable Energy 2013) in California involving 2,039 EV owners and drivers who had been driving EVs for more than six months, the most important reasons for switching to EVs were the environmental benefits and energy independence (figure 3.2). In Bhutan, however, these aspects are likely not the same because of the country's different sociodemographics, although additional information is required to make a full assessment (see also box 8.1 for a first brief analysis of reasons for early adopters in Bhutan to buy an EV).

Box 3.1 First Assessment: EV Awareness

When interviewing a number of Bhutanese drivers (a nonrepresentative sample), the study team learned that the drivers generally are interested in EVs but tend to emphasize the downside of electric vehicles, such as their limited driving range, battery degradation, and battery replacement costs. Opinions, however, were partly based on unreliable or nonexistent information. To improve EV awareness, a fact-based communication campaign from an independent source, such as an EV information center in Thimphu, would be very useful. In addition, an international seminar could be organized in Bhutan, involving regional and international EV experts as well as Bhutanese stakeholders, to improve knowledge and awareness.

Figure 3.2 Motivation for Electric Vehicle Purchases in California

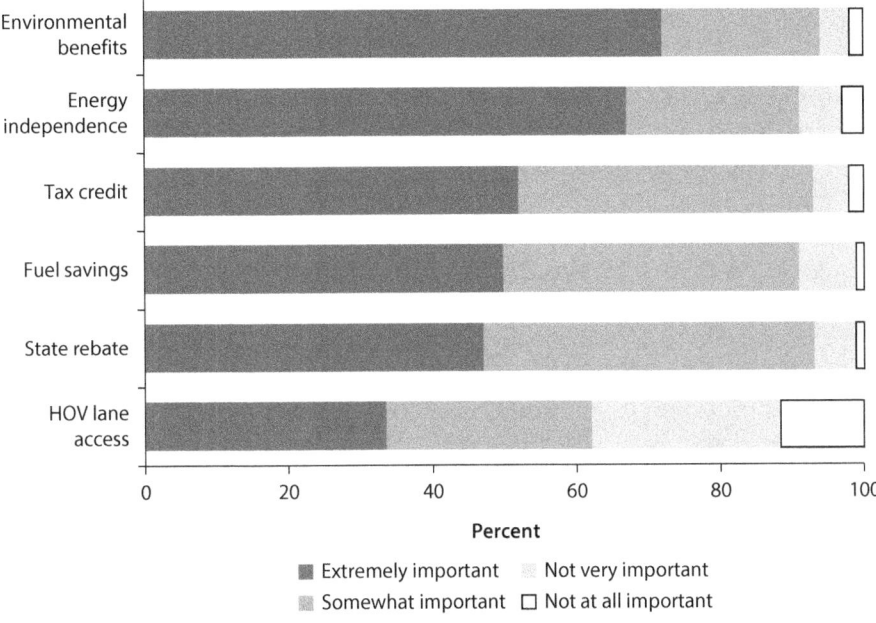

Source: California Center for Sustainable Energy 2013.
Note: HOV = high-occupancy vehicle.

Potential Market Segments in Bhutan

For the purposes of this study, four different target groups have been defined. The groups are based on international experience, which has shown that the first logical target groups to switch to EVs are government fleets, private vehicles, and vehicles in the taxi branch. In addition, public transport might, in time, switch to the use of EVs.

This report focuses only on passenger and light commercial vehicles and does not consider EV replacements for heavy trucks and two-wheelers. This is because few EV heavy trucks exist worldwide (mostly for demonstration purposes), and they have a very high development cost. Two-wheelers are not included because they will have little impact on fuel imports. They would also require a different charging infrastructure, have a different price range, and often involve different driving patterns.

Target Group 0: Public Transport

Public transport is a possible future target group, with conventional fuel buses switching to electric buses. The development of the electric bus, however, is in a more nascent stage than the development of EVs, even though many pilot projects are already testing the electric bus battery and charging technology, user experience, TCO, and bus operations. With increased demand for electric buses in metropolitan areas, technologies and prices are expected to develop quickly (Civitas 2013).

In Bhutan specifically, a large-scale deployment of electric buses would at the moment require very high capital expenditure. However, the potential to use simpler and lower-cost EVs to extend the coverage of public transport appears more promising. In Thimphu, for example, small-scale electric public transport vehicles could be introduced as a feeder mode to the city bus service, to extend the coverage of public transport services to districts of Thimphu where the financial viability of larger city buses is less viable. Nepal, for example, has achieved relative success with using simple, small-scale, inexpensive electric vehicles in its public transport fleet. In Kathmandu, about 600 electric "Safa Tempos" vehicles carry an estimated 127,000 people per day, with passengers paying fares ranging from NPR 15 to NPR 19 (US$0.15–0.20). Historically, the manufacturing price of each vehicle was in the order of US$5,000–7,000, and the vehicles at one time were assembled in Nepal. The drivers of the vehicles in Nepal own two sets of batteries per vehicle and swap charged sets at small private garages that operate throughout the city. A charged battery set delivers approximately 60 km of range. Drivers report that battery performance degrades after about 18 months, at which time they purchase replacements and sell old battery sets for scrap.

Target Group 1: Government Fleet

In many countries where EV use has grown in recent years, the governmental fleet has been the first target group. By switching the government fleet to EVs, the government supports the development of the market and shows that EVs are suitable for regular use, thus removing prejudices and making it more likely that companies and individuals will follow. As the government has control over its own procurement processes, it is relatively easy to include EVs in the fleet when budget is available. Moreover, charging and parking locations can be organized at government offices; the average daily distances traveled with most government vehicles is less than 100 kilometers per day, which fits well with the use of EVs.

Government vehicles that can most easily be transitioned to EVs are those that have only a limited operating area and do not carry heavy loads. This includes vehicles used for commutes to work, as well as vehicles for local inspection teams, courier services, local police, cleaning, and light garbage collection. Box 3.2 provides a first assessment of the government fleet as a possible target group.

Target Group 2: Private Cars

The private fleet is the largest segment of vehicles in Bhutan, but providing incentives to encourage this particular target group to switch from ICE vehicles to EVs can be difficult. Although the government can promote EV incentive programs, it has no control over the decision-making process of individual buyers. For this target group in particular, it is important that potential customers have the right information about the vehicles (such as the range of the vehicle, the TCO, and information about the degradation of the battery) before they decide

Box 3.2 First Assessment: Government Fleet as a Target Group

When switching the government fleet to EVs, it can be useful to start with a highly visible but small fleet that is used within the city area. This would minimize any problems related to the limited range of EVs. It will also be important to provide clear instructions to the drivers and provide reliable charging solutions to ensure a fair trial of the technology in the context of Bhutan. A sound evaluation scheme should be in place to capture good and bad practices related to the program and monitor and record driver experiences, charging behavior, and actual costs; experiences with the pilot fleet can then be communicated with other government organizations. Government-related companies, in particular those with possible business relations to the upcoming market of EVs—such as Bhutan Power Corporation (BPC), the Road Safety and Transport Authority (RSTA), and the Gross National Happiness Commission (GNHC)—could also procure EVs to gain experience in the EV market and the new value chain.

Box 3.3 First Assessment: Target Group Private Vehicles

The first group of pioneers who have bought a private EV in 2014 can be excellent ambassadors of EV driving in Bhutan. The group could be organized as an EV drivers club or association to channel and communicate experiences. The experience from these drivers and the association will offer valuable input for policy makers.

---

to buy. Without incentives, the number of private vehicle buyers who will voluntarily switch to EV will be small. Box 3.3 provides a first assessment of private vehicles as a possible target group.

Target Group 3: Taxi Fleet

The taxi fleet is an attractive market for EVs because of their high mileage and large cost-saving potential given Bhutan's low electricity costs. Taxis in Thimphu are often used for short distances, which also works well with the capacity of EVs. The high daily mileage of taxis, however, is a large disadvantage because of the need to charge the vehicle's battery several times a day. In addition, luggage space and loading capacity could be a problem, depending on the vehicle type and client.

A test with electric taxis in the Netherlands indicated that vehicles should have a minimum electric range of 200 km to fit into existing operations. In Shenzhen, where 50 electric BYD (Build Your Dreams) taxis with a range of 300 km are used,[3] both drivers and passengers are satisfied with the driving experience, although investment costs are fairly high because the electric taxi vehicles cost 80 percent more than the conventional taxi vehicles used (Yan and Durfee 2011). In Osaka, Japan, a fleet of 50 Nissan Leafs was introduced as electric taxi vehicles. The project (Japan Today 2013) reports positive customer

> **Box 3.4 First Assessment: Taxi Vehicles as a Target Group**
>
> The uptake of electric taxis worldwide is still limited, although several successful EV taxi projects exist internationally (see, for example, box 4.1). In Bhutan, the government could start to monitor, for one to two years, the driving and charging behaviors of the EV taxis already in use on the route from Paro to Thimphu. Based on the results, appropriate policies and incentive measures could be developed.

experiences, a smooth driving experience, and also a good exposure for EVs. However, the degradation of the battery, the longer charging time after two years of driving, and a negative business case due to the inability to make longer and more profitable trips are some of the less positive outcomes. In general, taxi markets are complex in terms of their political economy, which makes any predictions for this market segment difficult.

One potentially interesting market in Bhutan could be the high-end tourism taxi market. Providing sustainable transport fits well with the "image and experience" of the country and the approach of the GNHC. Although the costs of investment for suitable electric taxi vehicles (for example the Tesla Model S or Nissan Leaf) are often higher than comparable ICE vehicles, the higher investment costs can be recovered by using the taxis for high-end tourists who are willing to pay more for the service. A first logical testing area for EV taxis would be the transit from Paro Airport to Thimphu, which could be used to test usage and acceptance of EVs by taxi drivers. A few EV taxis already are active on this route. Box 3.4 provides a first assessment of taxi vehicles as a possible target group.

Three Scenarios for EV Uptake in Bhutan

To better understand the impact of an EV program in Bhutan, three scenarios have been developed assuming a respectively low, high, or super high uptake of EVs among the target groups. While the uptake rates are only a theoretical calculation based on certain assumptions, results can be used to estimate the impacts of such uptake rates in terms of required financial investments, economic impact, and charging infrastructure needs and costs. Each scenario considers the three main target groups—government fleet, private vehicles, and taxi vehicles—and is based on an estimate of the overall increase in vehicles for these target groups between 2015 and 2020 along with an estimated EV replacement rate.

Table 3.1 illustrates the estimated total number of vehicles in Bhutan for 2015 and 2020. The numbers for 2020 are based on those in 2015, using different growth rates for the three target groups based on current growth rates and policies. The overall growth rate for the number of taxi vehicles (23 percent) is an extrapolation of the current growth rates in this branch, while a slightly slower

Table 3.1 Estimated Total Number of Vehicles in Bhutan, by Target Group

Total number of vehicles	2015	2020	Assumption for estimated growth rate until 2020, %	Average annual vehicle replacement rate
Taxi fleet	5,271	6,500	23	736
Government fleet	2,524	3,000	19	345
Private fleet	36,736	49,379	34	5,382
Total	44,531	58,879		

Source: RSTA 2012–2013/World Bank analysis.

growth rate (19 percent) is used for the government fleet. The estimates have not considered specific conditions that could influence these growth rates, such as a limitation of the maximum number of taxis or increased efficiency in government vehicles. The growth rate for the private fleet (34 percent) is the highest and is also in line with current growth rates. Although the calculations use the total number of vehicles in Bhutan, most EVs are expected to be found in the Thimphu area. A detailed assessment per city or area might be done at a later stage.

Based on the total number of vehicles in 2015 and 2020, for each scenario a replacement rate (for all vehicles) can be calculated. The calculations assume the average lifetime of vehicles in the fleet is eight years. The overall regular replacement rate for 2015–2020 is then calculated as the average of both the 2015 and 2020 replacement rates. For example, for taxis, the regular replacement rate in 2015 is calculated as 5,271 vehicles/8 years = 659 vehicles per year in 2015, while that same replacement rate in 2020 is 6,500/8 = 813 vehicles. This means that, on average, 736 vehicles ((813 + 659)/2) are replaced each year for the six years from 2015 to 2020. For the government fleet and private fleet each year, on average, 345 and 5,382 vehicles are replaced.

Finally, using the calculated vehicle replacement rates, an EV replacement rate is applied for each scenario to calculate how many of the replaced vehicles will be EVs, with EV replacement rates based on international experience. In IEA EV countries, for example, EVs represent an average of 0.61 percent of new vehicle sales. For all three scenarios it is assumed that the growth rate of EV uptake will be linear from 2015 to 2020 (six years). This is assuming stable conditions like oil prices, battery prices, availability of EV models and incentive programs. Based on the total replacement rate of vehicles and EV replacement rate among new vehicles, a theoretical number of new EVs sold per year for each target group can be calculated.

Scenario 1: Low Uptake

For the low uptake scenario, EV replacement rates of 1 percent and 2 percent were assumed for the three target groups. The replacement rate of EVs for the government and private fleets is about 1 percent, which is just under the international replacement rate in places like the Netherlands and Norway. The EV replacement rate for taxis under this scenario is assumed to be 2 percent because this is the main target group for EV deployment in Bhutan.

Table 3.2 Number of Electric Vehicles by 2020, Low EV Uptake Scenario

	Taxi	Government	Private	Total
2015	15	3	54	72
2020	90	21	323	417

Source: World Bank analysis.
Note: EV = electric vehicle.

Table 3.3 Number of Electric Vehicles by 2020, High EV Uptake Scenario

	Taxi	Government	Private	Total
2015	74	10	161	245
2020	441	62	969	1,472

Source: World Bank analysis.
Note: EV = electric vehicle.

Table 3.4 Number of Electric Vehicles by 2020, Super High EV Uptake Scenario

	Taxi	Government	Private	Total
2015	736	17	269	1,022
2020	4,414	104	1,615	6,132

Source: World Bank analysis.
Note: EV = electric vehicle.

To calculate the total number of EVs by 2020, the EV replacement rates were combined with the overall vehicle replacement rates listed in table 3.1. For example, for taxis in the low uptake scenario (with a 2 percent EV replacement rate), each year 2 percent of 736 vehicles (or 15 vehicles), would be replaced by EVs. As a result, in 2020 there would be 6 years × 15 = 90 EV taxis on the road. Table 3.2 indicates the total number of expected EVs in 2015 and 2020 per target group in the low uptake scenario. The total number for private vehicles is high because this is by far the largest vehicle fleet segment.

Scenario 2: High Uptake
For this scenario, an EV replacement rate for government and private vehicles of 3 percent is assumed, which is higher than the replacement rate in countries with leading EV markets. For taxis, an even higher rate of 10 percent is used, meaning that 1 in 10 new taxis is replaced by an EV each year. Table 3.3 indicates the total number of expected EVs in 2015 and 2020 per target group in this scenario.

Scenario 3: Super High Uptake
The third scenario is calculated to match the RGoB ambition to introduce about 1,000 EVs per year. The EV replacement rate for taxis is estimated at 100 percent, which means that every new taxi is assumed to be an EV, while the EV replacement rate for the private and government fleets is assumed to be 5 percent (table 3.4).

Figure 3.3 Number of Electric Vehicles for the Three Uptake Scenarios, 2015–2020

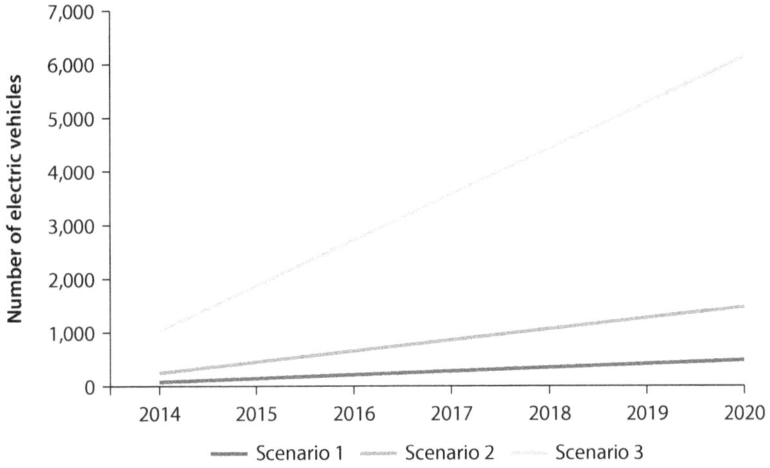

Source: World Bank analysis.

Figure 3.3 presents an overview of the total estimated number of EVs for the three scenarios based on the calculations. Actual numbers, however, will depend on various factors, such as incentive programs, petrol price, and consumer acceptance of EV (see also chapter 5).

Notes

1. The International Council on Clean Transportation (ICCT), on low relative EV market share in Denmark.
2. This report is based on international experience with EV, presented in the context of Bhutan. Possible first actions, based on an initial assessment by the project task team, are indicated in a set of "First Assessment" boxes throughout the report.
3. Jiangnan taxi division, Shenzhen.

References

Anegawa, Takafumi. 2010. "Characteristics of CHAdeMO Quick Charging System." Paper prepared for the EVS25 World Battery Hybrid and Fuel Cell Electric Vehicle Symposium, Shenzhen, China, November 5–9.

California Center for Sustainable Energy. 2013. *California Plug-in Electric Vehicle Driver Survey Results*. May 2013.

Civitas. 2013. *Smart Choices for Cities—Clean Buses for Your City*. Civitas.

Hannisdahl, O.H., H.V. Malvik, and G.B. Wensaas. 2013. "The Future Is Electric! The EV Revolution in Norway—Explanations and Lessons Learned." Paper prepared for the Electric Vehicle Symposium and Exhibition (EVS27), Barcelona, Spain, November 17–20.

Hjorthol, R. 2013. *Attitudes, Ownership and Use of Electric Vehicles—A Review of Literature*. Oslo, Norway: Institute of Transport Economics.

Japan Today. 2013. "Osaka's Great EV Taxi Experiment Does a Slow Burnout." February 20.

National Statistics Bureau and Asian Development Bank. 2012. *Bhutan Living Standard Survey*. Thimpu.

Yan, Fang, and Don Durfee. 2011. "Analysis: Chinese Electric Taxis Struggle to Win Mass Appeal." Reuters. June 29.

CHAPTER 4

Electric Vehicle Market and Technology Development

Key Messages

- Globally, the electric vehicle (EV) market is still in a development phase with a penetration of about 0.02 percent of the total number of vehicles.
- The average expected range of all EVs in Bhutan is 50–150 kilometers. Factors influencing range include the size and effectiveness of the battery, outside temperature, driving style, and topography.
- Electric range is expected to grow in coming years with the development of lithium-ion batteries with a high energy density.
- A policy for the second life of EV batteries needs to be developed.

Global EV Market Development

In recent years, the number and variety of EVs has increased significantly; almost every original equipment manufacturer (OEM) has developed at least one EV model. By 2020, about 20 million EVs are expected to be on the road worldwide, with most cars driving in China, Europe, and the United States (figure 4.1) (Trigg and Telleen 2013).

Because EVs are an innovation, the diffusion of innovations curve in figure 4.2 can be used to analyze their market penetration (Rogers 2003). The first buyers—the innovators—are a small group representing about 2.5 percent of the market share of EVs. Considering global sales numbers, the uptake of EVs worldwide is still on the far left side of the innovations curve.

Figure 4.3 provides an overview of the EV market share in several countries. Despite the growing numbers of EVs, market shares are still small in most countries. The main reasons for the relatively large market shares in the Netherlands and Norway are the extensive fiscal and policy incentives the governments have put in place (see also chapter 5). Since the global market is still in an early adoption phase, countries with a large EV market share have high costs in terms of fiscal incentives and infrastructure investments; this means that

Figure 4.1 Electric Vehicle Uptake Worldwide

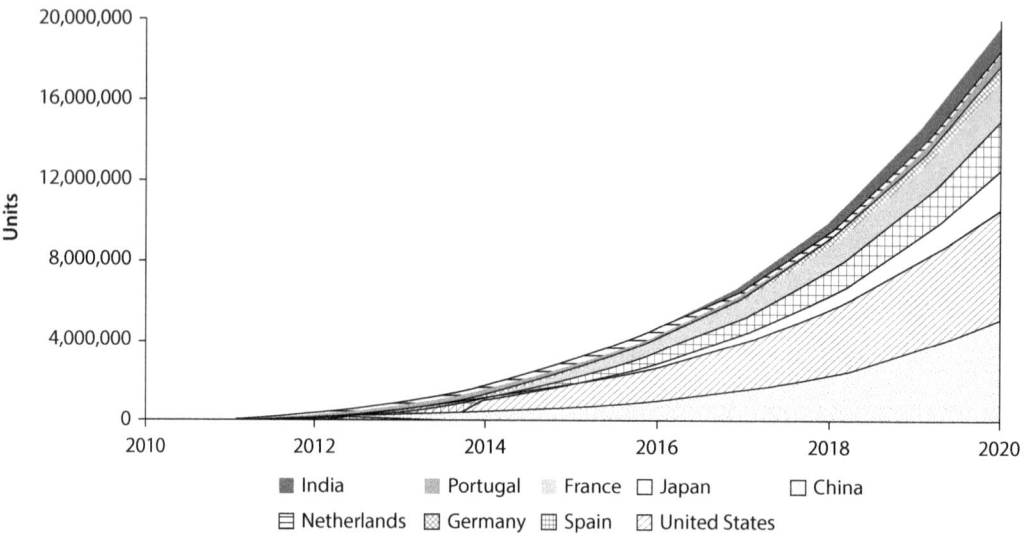

Source: Trigg and Telleen 2013.

Figure 4.2 Diffusion of Innovations Curve

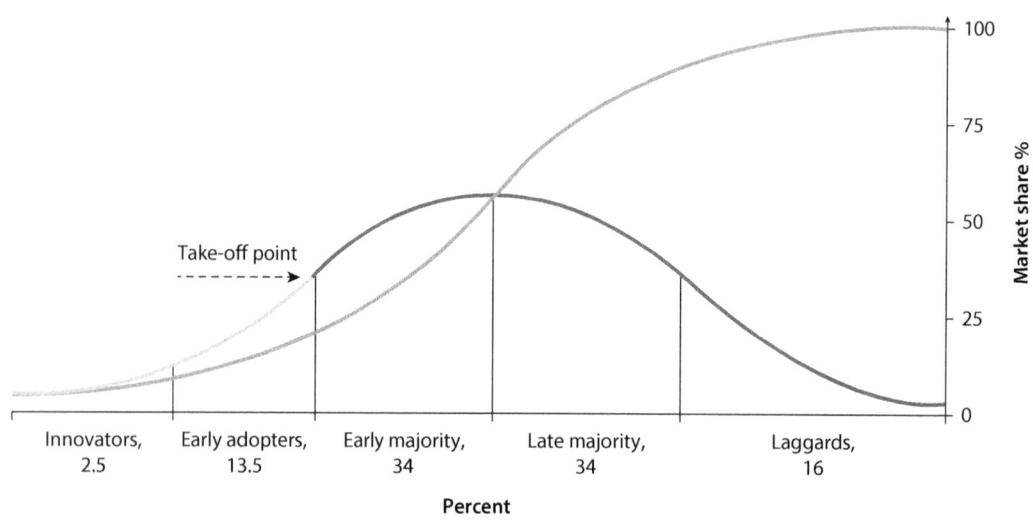

Source: Rogers 2003.
Note: Globally, the market is still in the "**innovator**" phase, with a **penetration** of about **0.02%**.

for countries without the capacity for large financial investments, a moderate approach might be more fitting. Adopting a more gradual uptake strategy is also likely to be more effective as technology advancements will continue to make it easier for drivers to switch to EVs, reducing the need for fiscal support to encourage EV purchases.

Figure 4.3 Electric Vehicle Market Share per Country

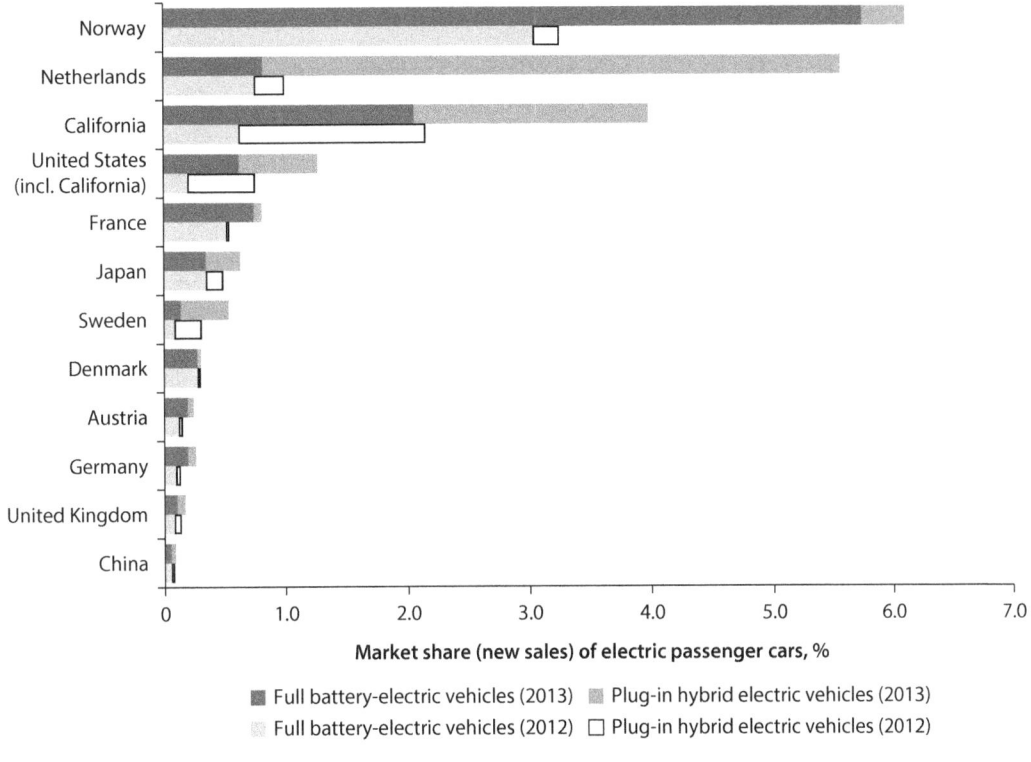

Source: Mock and Yang 2014.

In general, sales and production of EVs are expected to grow rapidly in coming years, with significant growth expected for Asia and in particular India and China (Kalmbach et al. 2011). These developments could have a positive impact on the EV sector in Bhutan because they will lead to more affordable EVs and possible industrial opportunities in the country. The charging standards and technology that will be chosen by China and India are also expected to have a major influence on Bhutan's EV sector. For China, this decision will be made in 2015; for India, it is expected in 2016.

Types of EVs: Plug-In Hybrids and Full Electric Vehicles

Many different types of EVs exist—all with different specifications for the drive train technology, electric range, and driving characteristics. In general, EVs can be divided into two main categories, both of which are considered for this report:

- **Full electric vehicles (FEVs).** FEVs drive only on the battery and have an electric range that varies between roughly 80 km (for the Mahindra) and 400 km (Tesla Model S). The average range for the industry is about 100–150 km. The average energy in kilowatt-hours (kWh) required for each kilometer for FEVs is about 0.2, but this depends on the vehicle and manufacturer.

- **Plug-in hybrid electric vehicles (PHEVs).** PHEVs have a battery with an electric drive train as well as an internal combustion engine (ICE). The all-electric range is roughly 40–60 km. When the battery is empty, the ICE can recharge the battery or serve as a backup to drive the vehicle.

Technology development for both PHEVs and FEVs has seen only a first wave of investment by most OEMs, resulting in a few first models. As markets expand, more research and development (R&D) will be needed. In the near future, the market share for PHEVs is expected to grow much faster than that for FEVs, mainly because of better customer acceptance and product versatility. Incentive programs, however, can influence this PHEV/FEV ratio.

At the moment, the Nissan Leaf is the best-selling FEV worldwide. With over a 100,000 vehicles sold, it is far ahead of other popular models such as the Tesla Model S, Mahindra e2o, Mitsubishi I-MiEV, and the BMW i3. In Bhutan a number of Nissan Leafs are also already on the road, along with the Mahindra and Tesla models. The Nissan Leaf has a 24 kWh lithium-ion battery in 48 lithium-ion modules that store its energy to power an 80 kW alternating current (AC) motor. According to the New European Driving Cycle (NEDC), the Leaf has a range of 200 km, while the Japanese standard test cycle (JC08) gave it a range of 228 km (see also "Factors Influencing Driving Range," below). A second-generation Leaf is expected for 2016–2017, with more battery capacity and more range.

Factors Influencing Driving Range

Driving range is an important factor for people buying an EV, and this range is influenced by various characteristics of the vehicle as well as outside conditions. Although the different types of EV all have an official range (for example, the Nissan Leaf has an NEDC driving range of 200 kilometers), this is only a theoretical number determined in a test of the vehicle under ideal conditions. International experience and studies have shown that actual driving range depends on the characteristics of the vehicle, such as the size and effectiveness of the battery, the effectiveness of the drive train, and aerodynamics and tires. In addition, outside temperature, driving style, speed, topography, and the use of heating or air conditioning play a role. The circumstances and range of an "average ride" in Bhutan will be discussed in "An 'Average Ride' in Bhutan," below.

Battery Size and Effectiveness

EVs currently on the market vary in battery capacity and range (table 4.1). In the case of the Nissan Leaf, which has an NEDC range of 200 km, average range is actually only about 120–150 km per drive. The Mahindra has a range of 50–80 km per drive, while Tesla has the largest average range of 300–500 km. The range of the Tesla, however, is not representative of the industry average, although this could change because Tesla has given away its patent to speed up the development of EV technology.[1]

Because the battery is one of the most expensive components of the car, a battery lease is one way to reduce the consumer risk of a degraded battery. Battery performance and lease are further discussed in "Battery Performance and Battery Second Life," below.

Aerodynamics and Tires

The power consumption of the car is greatly influenced by its aerodynamics and tires, as shown in figure 4.4. Several EVs, such as the egg-shaped Toyota Prius Plug-in, Nissan Leaf, and Tesla Model S, are relatively effective in minimizing the friction with both air and road. The Tesla, for example, has "air ride suspension," which lowers the car about an eighth to a tenth of an inch during highway speeds. Wind also is a major influence on range. High wind speeds

Table 4.1 Battery Capacity and Range of Selected Electric Vehicles

EV type	Battery capacity	Average electric range (km)
Nissan Leaf (FEV)	24 kWh	120–150
Mahindra e2o (FEV)	19 kWh	50–80
Tesla Model S (FEV)	60/85 kWh	300–500
Mitsubishi i-MiEV (FEV)	16 kWh	90–120
BMW i3 (FEV)	22 kWh	120–140
Opel Ampera (PHEV)	12 kWh	40–60
Mitsubishi Outlander (PHEV)	12 kWh	30–50

Source: World Bank analysis.
Note: Electric range is based on user experience and driving on a flat terrain. FEV = full electric vehicle; kWh = kilowatt-hour; PHEV = plug-in hybrid electric vehicle.

Figure 4.4 Factors Influencing Electric Vehicle Power Consumption

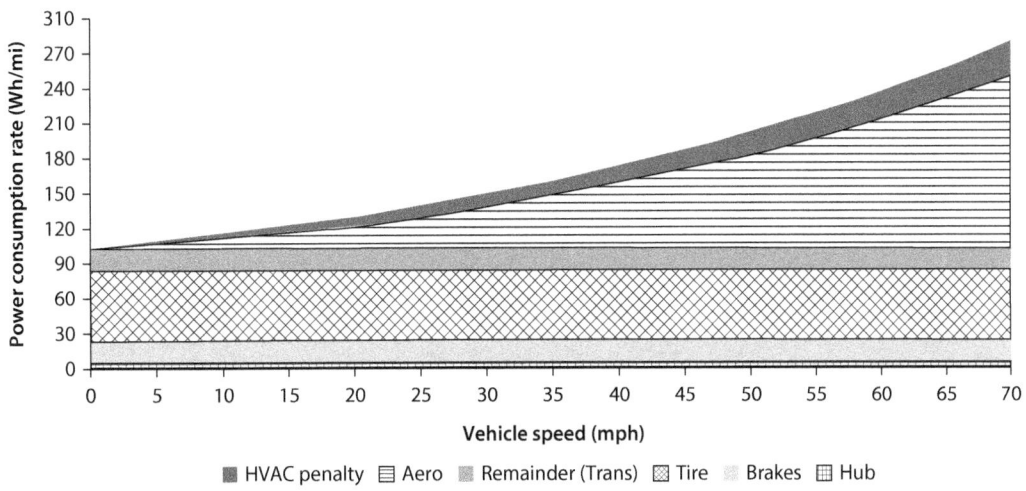

Source: www.teslamotors.com/goelectric/efficiency.
Note: HVAC = heating, ventilating, and air conditioning; MPH = miles per hour; Wh/mi = watt hour/mile.

countering the vehicle can cause up to 15 percent range loss, whereas tailwinds will render comparable extra range.

Outside Temperature

In addition to vehicle characteristics, other circumstances—such as outside temperature—also influence range. Figure 4.5 shows the impact of temperature on the range of a Nissan Leaf. On average, the most advantageous outside temperature for battery performance is about 15–25 degrees Celsius. Both lower and higher temperatures will lower battery volume and range. As shown in the figure, at about 0 degrees Celsius, the range is decreased by about 20 km compared to a drive at optimum outside temperatures.

Driving Style

As with ICE vehicles, the performance and range of an EV are influenced by driving style: for example, sudden acceleration causes energy and range loss. Only part of the energy that was used for the acceleration can be regenerated when decelerating or breaking, and EV performance is at its best at a constant speed. Urban trips, because of their frequent starts and stops, will therefore typically use more energy than rural trips with the same average speed. The use of a vehicle's air-conditioning (A/C) system will also impact driving range because it consumes a lot of energy from the battery.

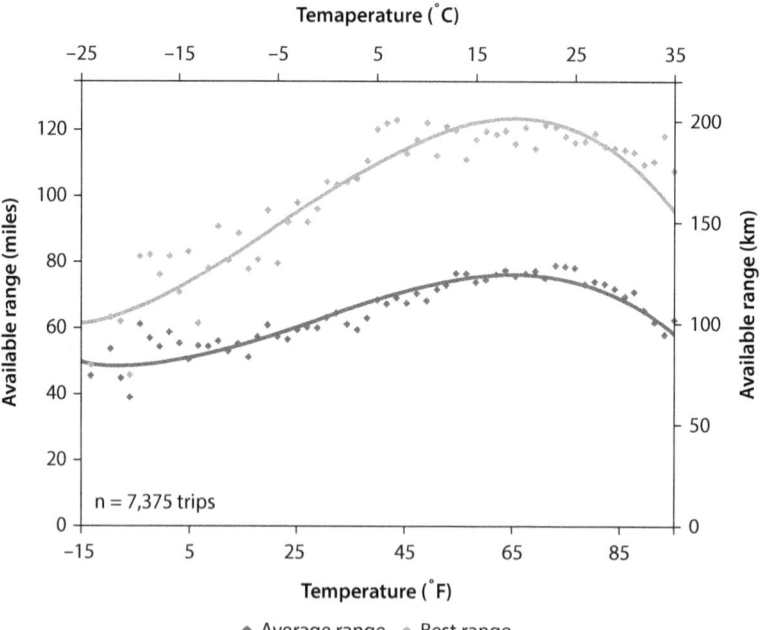

Figure 4.5 Nissan Leaf: Range vs. Temperature

Source: FleetCarma monitoring, December 2013.
Note: In **Thimphu**, the average **temperature** is **about** 20 degrees **Celsius** (May–September). The average range of a Nissan **Leaf** is **about 100** km (corrected for mountainous area).

Speed

Driving speed, even more than driving style, also strongly influences range. Best performance has been measured at a constant velocity of 15–40 km/h. At high speeds, such as on motorways, range drops to as little as half of the possible maximum range. Figure 4.6 illustrates how the range of the Tesla Model S depends on speed.

Topography

The topography and profile of the area where the EV is driven also influence the range of the vehicle. When the terrain is relatively flat, range is not affected. In a more mountainous area, however, going up will significantly reduce range. When driving between 40 and 120 km/h, energy consumption can increase up to ninefold when the slope changes from 0 to 20 percent (Moure, Roche, and Mammetti 2013). Similarly, on a downward slope of 0–20 percent, energy consumption can decrease by a factor of three. On average, a slope increase of 1 percent results in an increase in energy consumption of 0.063 kW/km.

Figure 4.6 Tesla: Range vs. Speed

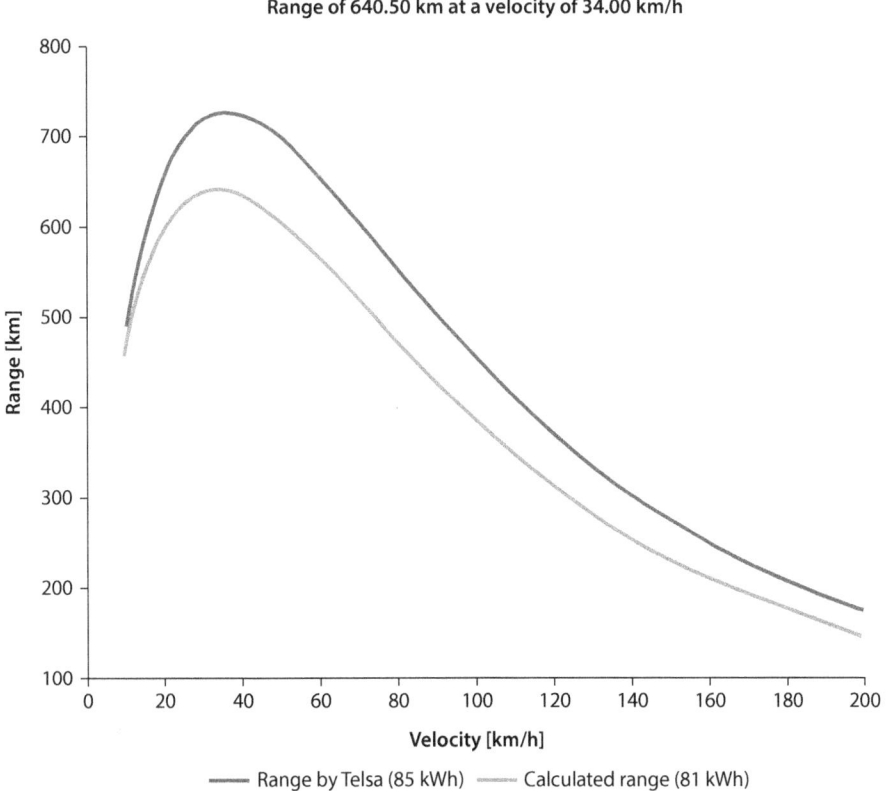

Source: Blogpost of Professor M. Steinbuch, Technical University Eindhoven, August 28, 2013.
Note: Graph based on computer-simulated data. kWh = kilowatt-hour.

An "Average Ride" in Bhutan

When combining the various vehicle characteristics and other factors influencing driving range (see "Factors Influencing Driving Range," above), it is possible to give an indicative range of an "average ride" of an EV in Bhutan. Even so, however, range would still depend on the specific conditions. For example, assuming a drive at an average speed of 50 km/h through a mountainous area, the range of a first-generation Nissan leaf would be about 100 km in summer, when the temperature is about 25 degrees Celsius in Thimphu.[2] In winter, however, with a night temperature of −2 degrees, the same car making the same drive would have a range of only about 75 km.[3]

Optimal conditions for driving the Nissan Leaf are 20 degrees, flat terrain, A/C turned off, and a cruising speed of about 30 km/h. Figure 4.7 illustrates how the predicted range of the Nissan Leaf depends on various factors. The second-generation Nissan Leaf is expected to offer more range as a result of improved battery technology and more efficient heating.

When also considering the terrain, the range may be further reduced. A one-way journey from Thimphu to Phuentsholing is 154 km. This downhill route (figure 4.8) might be possible with a full battery of a Nissan Leaf, assuming the A/C is off, only one passenger is in the vehicle, and the ambient temperature is optimal. A less optimal condition, however, would be the return trip uphill, in cold conditions (0 degrees Celsius), with four passengers and luggage, and with

Figure 4.7 Nissan Leaf: Range Predictor

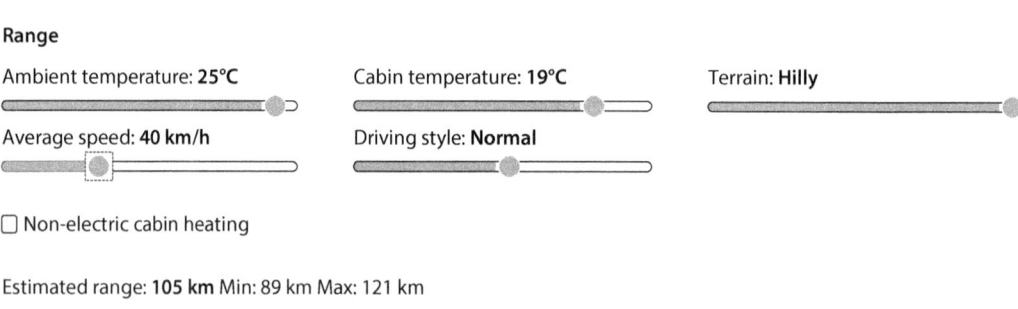

Estimated range: **105 km** Min: **89 km** Max: 121 km

Source: Grønn Bil Norway 2012, http://www.gronnbil.no/rekkevidde.

Figure 4.8 Elevation Profile of the Route Thimphu–Phuentsholing

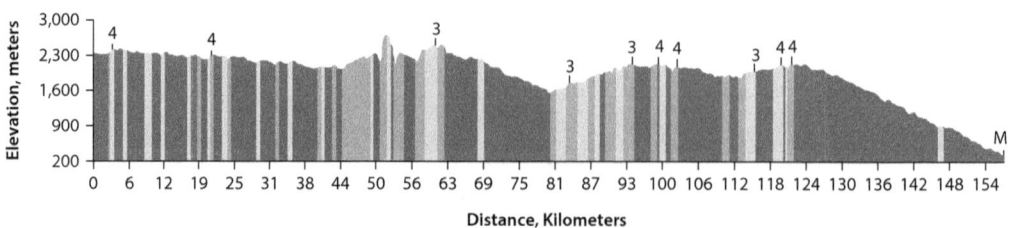

Source: Google Maps.

the A/C turned on, in a first-generation Nissan Leaf. Under these conditions, the indicative range is only about 60 km.[4]

When the range of the EV is not enough for a trip, fast charging can be used to extend it. Because a fast charger charges a battery to only 80 percent of its capacity, for the Phuentsholing–Thimphu journey at least two fast charging locations would be needed along the route. The driving time for the trip would then also increase by about one hour (2 charges x 30 minutes) to accommodate the charging. An average distance between fast chargers of about 40–60 km would be sufficient to serve most EVs.[5] Because the main road network of Bhutan is about 2,000 km,[6] approximately 50 fast charger locations would be needed for the country. Charging infrastructure is further discussed in chapter 6.

International User Experience with EVs

The performance of EVs has been monitored in several international projects, including the Switch EV Trial and the Green eMotion project. The Switch EV Trial, which took place over a period of 2.5 years from 2010 to 2013 (Hübner et al. 2013), was situated in England's northeast and consisted of 44 EVs, including the Nissan Leaf and the Peugeot iOn. Participants included a mixture of companies, local authorities, and private individuals. The trial showed, for example, that participants used their EVs mostly for trips that were under 40 km; only 2 percent of participants recorded a journey longer than that (figure 4.9). Because in the United Kingdom (as in the Netherlands) most trips (nearly 95 percent) with ICE vehicles are also under 40 km, the trial showed that EV usage was similar to that of normal cars. In addition, 40 km is well within the driving range of EVs.

During the trial, over 90,000 journeys and 19,000 charging events were recorded. Of all participants, 17 percent charged their EVs twice a day, although most of the journeys were less than 15 km, illustrating how *range anxiety* impacts

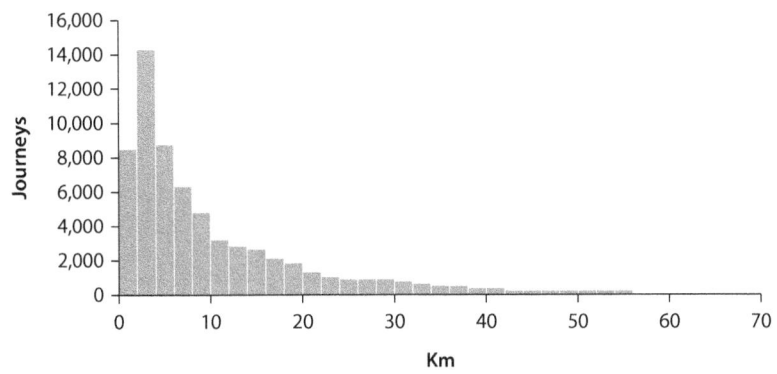

Figure 4.9 Number of Journeys and Average Distance

Source: Hübner et al. 2013.

charging behavior. Data from the trial also showed that both home and workplace charging are essential for the average EV driver (figure 4.10).

In the Green eMotion project, additional information was collected on the usage of EVs and charging stations (McDonald et al. 2012). The project was a pan-European, large-scale project aimed at researching and delivering a scalable framework for the rollout of EV technology. Data were collected from 10 demo projects in different regions throughout Europe. For the project, 235 EVs, 598 charging points, and 269 individual users were included in the monitoring and data collection.

The results of the project were similar to those of the Switch EV Trial. The project showed that home charging was responsible for the highest level of energy consumption, with an average of 5.9 kWh per charge. Most of the EVs were plugged in after arriving home from work between 5:30 p.m. and 8:30 p.m. Charging events in the weekend often showed a lower energy consumption (0–4 kWh per charge) than did weekday charging events. Although home charging represented the highest energy consumption, the most frequently used charging stations were office based. The usage level of public charging stations was the lowest of all locations, with an average time of two hours spent at those stations. In terms of traveling patterns, 90 percent of all recorded trips were less than 30 km, with an average of 8.7 km per trip.

International experience further suggests that specific target groups need additional charging facilities compared to regular EV users. While most target groups such as commuters and recreational drivers will be satisfied with

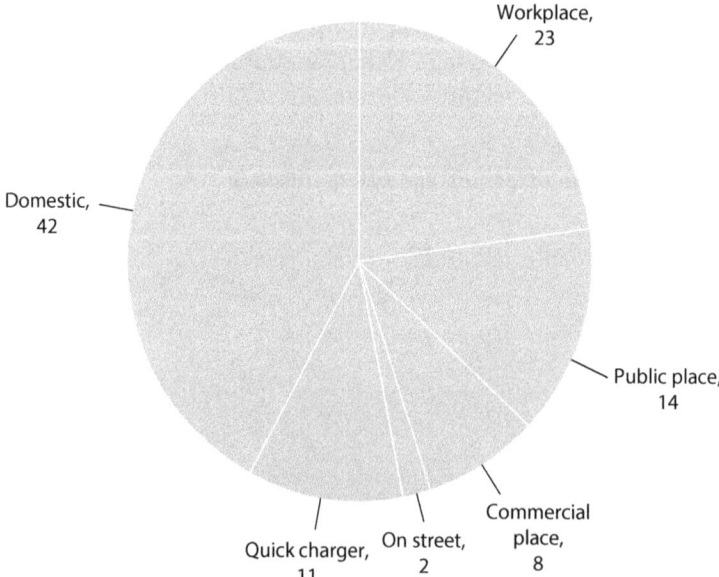

Figure 4.10 Charging Locations Used in the Switch EV Trial

Source: Hubner et al. 2013.

home and workplace charging, drivers who regularly use their FEVs for longer distances need additional charging options. This includes drivers who use EVs for their core business, such as taxi drivers, delivery services, and transportation companies. Examples of failed and successful EV taxi projects—both in the Netherlands—are described in box 4.1.

Box 4.1 Case Study: Experiences with EV Taxis in the Netherlands

In the Netherlands, two EV taxi projects have shown the difficulties involved with using an electric taxi fleet. With the help of some government support, however, one of the projects managed to become a successful business.

Utrecht taxi demonstration project. In Utrecht, a city of 300,000 inhabitants, an EV taxi demonstration project ran for only half a year until it was stopped because of a lack of operational effectiveness. The project, which received a total subsidy of US$2 million (Nu 120 million) from the national government, had been hindered the most by a lack of fast charging for the taxi fleet. Because the project ran in the very early stage of EV market development, not all taxi vehicles were suitable for fast charging and only a limited amount of suitable vehicles were available (the Nissan Leaf, for example, was not included). Moreover, fast charging was not combined with the charging stand and it was not certain if a direct current fast charger would be installed. Finally, public charging was difficult to set up as it took the city some time to develop a clear policy for electric driving, while the permit process for public charging stations was very slow. Also, according to the trial company, an average taxi should be able to drive at least 200 km per day on a single charge to be able to deliver the expected service and have a successful business case, but the vehicles were not able to do that.

Taxi Electric in Amsterdam. In a second project, in Amsterdam, a more successful experience with an all-electric taxi fleet was established. The fleet, consisting of Nissan Leafs, operated in the center and region of Amsterdam. The project was realized without restrictions from the authorities on conventional ICE taxis, and the taxi company, Taxi Electric, also received a subsidy of US$6,750 (Nu 405,000) per taxi from the municipality of Amsterdam, as well as an additional subsidy of the same size from the Netherlands government. Taxi Electric currently has about 25 taxis and this number is still growing. In September 2014, some Nissan eNV200s were included in the fleet, constituting the worldwide introduction of eNV200 as EV taxi. This larger vehicle provides more luggage space, for example for transportation to and from the airport.

The experiences from Taxi Electric have pointed to the importance of contracted and planned transport for the effective use of electric taxis, as this makes it possible to plan driving and charging times. The driving and charging behavior of the taxis is also constantly monitored for the project. Just outside the city center, a large charging hub (the largest of its kind in Europe) is used, which has four DC fast chargers and about 40 normal chargers. During the day, the taxis can be charged at the fast chargers, while overnight the normal chargers are used. Taxi Electric is also a social project, and actively takes on drivers from over age 50, for whom it is more difficult to find a job because of unemployment or other factors.

box continues next page

Box 4.1 Case Study: Experiences with EV Taxis in the Netherlands *(continued)*

Considering these and other international experiences with EV taxis, some considerations for new EV taxi projects might be to

- Start with a small number of vehicles on routes where the limited range of EVs does not directly influence the business case.
- Use relatively fixed route schedules and ensure the taxi company is well informed of operational changes to allow for the charging of EVs.
- Involve government subsidies to achieve comparable total cost of ownership (TCO).
- Ensure fast and slow charging are in place from day one.

Battery Performance and Battery Second Life

Battery Performance and Life Span

The battery of an EV is a critical element and determines a large part of the vehicle's characteristics. See figure 4.11 for an overview of the high-level composition of a battery. Currently, several types of batteries exist, which all present trade-offs in terms of cost, life span, performance, energy, power, and safety. The battery is also expensive. Various factors have an impact on battery life:

- **Battery type and chemistry.** Various combinations exist for battery chemistry (lead-acid, LiFePo4, Li-ion), all resulting in a different performance in terms of power, aging, and cost.
- **Number of charging cycles.** All batteries degrade a little with every charging cycle. The loss per cycle differs considerably per battery.
- **Charging/discharging depth.** A relatively small depth of discharge (DoD) is better for the battery. Charging and discharging depth are influenced by the (a) charging profile, (b) the ratio between slow and fast charging, and (c) driving behavior and loading:
  - *Charging profile.* In general, when the battery is used in the sweet spot of the battery capacity (20–70 percent), the total number of times a battery can be charged over its lifetime is higher than when a battery is used to capacity each time it is charged.
  - *Ratio between slow and fast charging.* Fast charging influences battery life. Although some specific EV batteries are designed for fast charging, they will in fact degrade slightly faster when charged with DC fast charging. In a comparison of two Nissan Leafs, with one vehicle using only slow charging and the other fast charging, after 50,000 km the type of charging had only a small effect on battery capacity with a maximum amount of 5 percent additional degradation (Shirk and Wishart 2015).
  - *Driving behavior and loading.* Both driving behavior and loading influence the discharging pattern of the battery.

Electric Vehicle Market and Technology Development 33

Figure 4.11 High-Level Composition of Battery

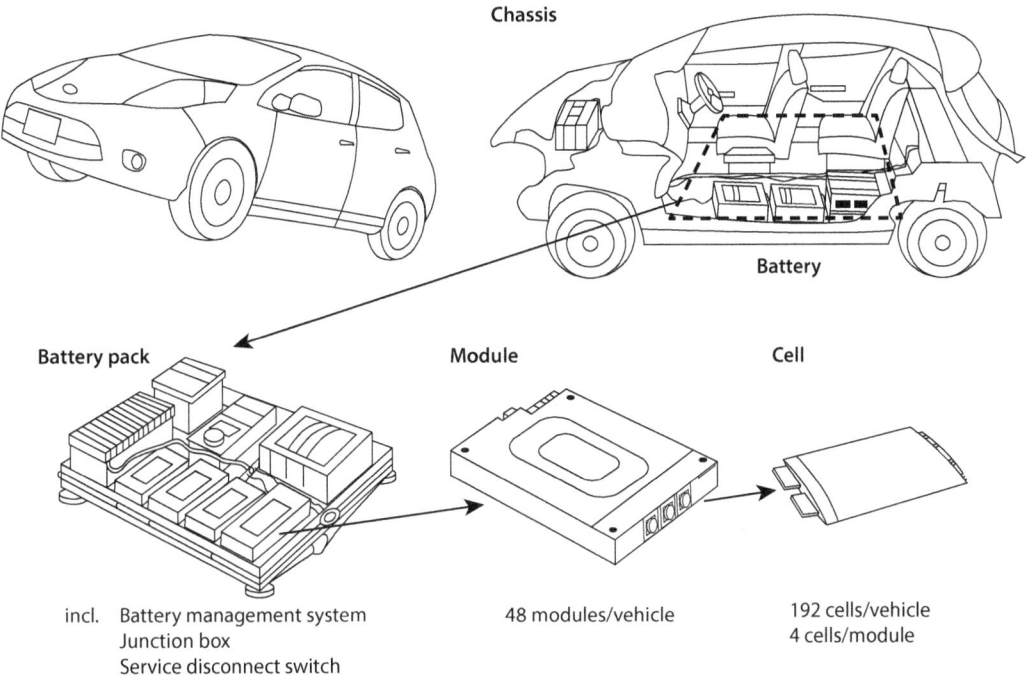

- **Temperature.** Extreme temperatures—either very high or very low—greatly influence the life span of the battery. Some vehicles have cooled battery packs for ambient temperature protection.
- **Time (aging).** Even without driving, battery quality decreases a little bit every year.
- **Battery management system (BMS).** A BMS is an electronic system used to manage a rechargeable battery (cell or battery pack), for example to protect the battery from operating outside its safe operating area (SOA). A BMS can also monitor the state of the battery, calculate and report secondary data, control its environment, and authenticate or balance the battery (Barsukov and Qian 2013).

Battery Types and Developments in Battery Technology

The energy density of batteries in weight (Watt hour per kilogram, Wh/kg) and volume (Watt hour per liter, Wh/l) is developing gradually. At the moment, rechargeable lithium-ion (Li-ion) batteries still seem to be the most practical solution for powering EVs. The main limitation is the size of the battery. Prototype cells seem to demonstrate that it is possible to develop next generations of Li-ion batteries with high energy density, with an impressive gain in energy density compared to conventional Li-ion batteries.

Other types of batteries are still in the developing stage, such as, for example, the aluminum–air cells that could deliver higher energy density (<400 Wh/kg) than primary batteries. Problems to overcome are the corrosion of aluminum under certain conditions and the formation of a passive hydroxide layer on aluminum, which inhibits dissolution and shifts its potential to positive values. Different solutions are being developed taking into account the most cost-effective solutions, since high purity aluminum is expensive (Egan et al. 2013).

Another alternative is the zinc–air battery, which is commercially available as a primary battery but for which recharging has remained elusive (Siahrostami et al. 2013). Rechargeable zinc-air batteries could be ideal with a high energy and power density, as well as high safety and economic viability.

The development of a new battery system—from electrochemical reactions in the lab to reliable working devices that can be produced on a mass scale—may require many years of sustained R&D efforts followed by time-consuming engineering work and tests before production lines can be set up (Yoo et al. 2014). The accelerated production of EVs and the need for large-scale storage of sustainable energy will boost the development of rechargeable batteries, but there is still a long way to go in the innovation curve (Yoo et al. 2014).

The expectation is that lithium-ion will remain the standard in the coming years (see figure 4.12 for an overview of the battery composition). This is emphasized by the fact that the "gigafactory" that Tesla plans to build and start production in 2020 will focus on producing lithium-ion batteries. With the production

Figure 4.12 Battery Composition of a Lithium Iron Phosphate Battery (LFP Battery)

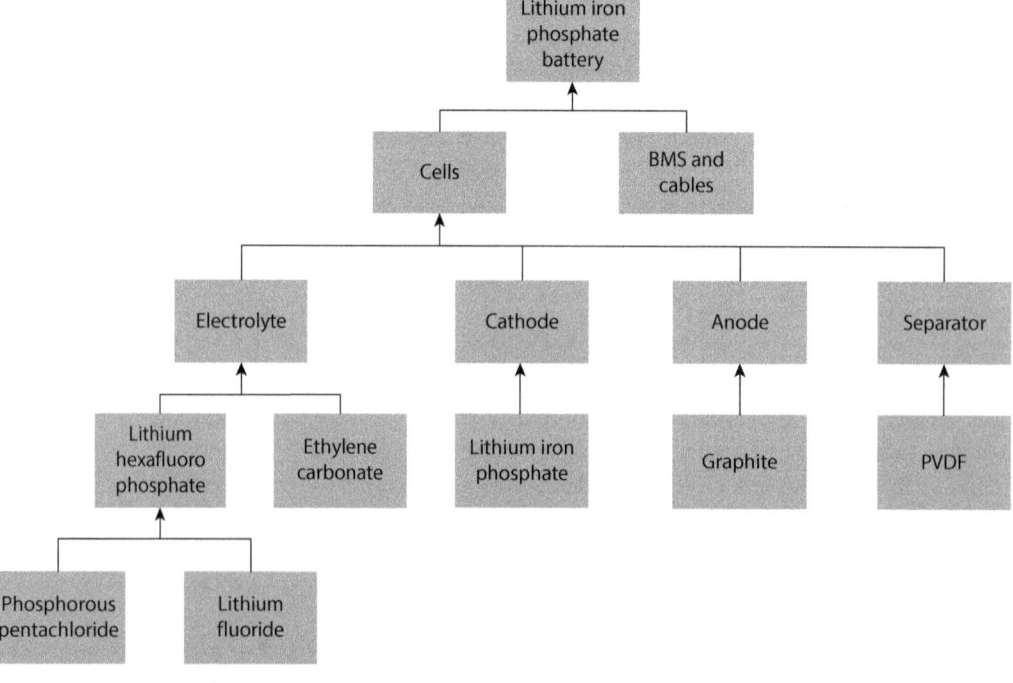

Source: World Bank.

of batteries in this gigafactory, the price can be lowered to US$100 (Nu 6,000) per kWh (instead of the current price of US$300–500 [Nu 18,000–30,000] per kWh), which makes it more compatible to the price of ICE vehicles (The Economist 2014). In terms of compatible range, only Tesla has managed to create a range similar to ICE vehicles at the moment. The main reason for this is that the capacity of the battery is much bigger (65–80 kWh) than that of other OEMs such as Nissan (20 kWh).

Manufacturer Battery Warranty

To address consumer concerns about battery life, EV manufacturers offer battery warranties. Conditions vary, as shown in table 4.2.

Another option offered by manufacturers is the battery lease. In this case the vehicle is purchased and the battery is leased for a fixed price over a fixed period (for example 24 months) and a fixed number of kilometers (such as 10,000 km/year). When buying an electric car from Renault (such as the Kangoo Z.E. or Zoe), for example, consumers are required to lease the battery for about US$90–150 (Nu 5,400–9,000) per month, depending on their average driving per year and the number of years that the car has been used. With Nissan, consumers have the option to either buy or lease the battery. Leasing the battery greatly reduces upfront costs and the risk of battery life degradation. Disadvantages of a battery lease are the low consumer acceptance of split ownership and the legal and financial challenges. Battery leases also influence financial incentives because the initial cost of the car will be lower.

Battery Second Life

As described above, the quality of the battery decreases gradually as a result of several factors. A vehicle, however, can still be used for many years even when the range is less than the original range. If for instance a private vehicle is driven 10,000 km per year with a maximum of 50 km per day (a very common maximum per day), a vehicle with a degraded battery will still be suitable for at least 10 years, even when the battery capacity is down to 70 percent of its original capacity. The battery in this case might last as long as the vehicle itself. Driving 100,000 km per year, however, advances the need for battery replacement because the capacity of a degraded battery will not be enough for this intended use.

Table 4.2 Original Equipment Manufacturer Battery Warranties

Manufacturer	Warranty	Details
Nissan United States	8 years or 100,000 miles	A loss of 3 out of 12 bars is within warranty, no formal fast charging limit
Tesla P85	8 years, unlimited mileage	Covers 70% of capacity
Mahindra Reva	3 years and 60,000 km	Covers 50% of capacity
BMW United States	8 years or 100,000 miles	Covers 70% of capacity, unlimited fast charging

Source: Manufacturers' information.

When the battery can no longer be used for its EV, the battery's "second life" begins, and a policy might be needed to guide this second life of EV batteries. Suitable options for Bhutan include a return of the battery or vehicle to the OEM, reuse of the complete battery pack, or recycling of the battery:

- **Return to OEM.** It could be a norm that every EV imported to Bhutan has a guaranteed return policy to the OEM or importing company. For example, Nissan has informally indicated that in the United States a battery replacement for the Nissan Leaf would cost US$5,500 (Nu 3,300,000), including a US$1,000 (Nu 60,000) refund for the old battery. The replacement pack would have the same warranties as a new vehicle (eight years or 100,000 miles).

- **Reuse of the complete battery pack.** The complete battery packs that are used in EVs have a great capability to store energy and could be used to store renewable energy, such as solar or small-scale hydropower, to balance the periods when energy is produced and used in households or apartment buildings. The complete battery packs could also be used at home for power outages. A Nissan battery with 50 percent degradation can still power an average household for three continuous days or a four-family apartment building for half a day.

- **Battery recycling.** A battery includes various components such as modules, cells, and raw materials, which (at different scales) could be reused. Currently, however, battery recycling is still far more expensive than sourcing new materials because of the sufficient availability of raw materials, the limited number of recycling facilities, and the need for labor-intensive activities and special care for hazardous and toxic materials.

The recycling of batteries is also important for ICE vehicles, which use small lead-acid batteries. As these lead-acid batteries require more raw material than Li-ion batteries to achieve the same energy storage, they actually have a much larger impact on the environment as a result of the raw material mining. Because of their frequent use, however, the recycling process for lead-acid batteries is well developed in many countries.

Notes

1. http://news.discovery.com/autos/fuel-and-alternative-fuel-technologies/tesla-gives-up-patents-to-spur-innovaiton-140613.htm.
2. http://www.tourism.gov.bt/about-bhutan/climatic.
3. EV range predictor from Norway used for Nissan LEAF MK1 http://www.gronnbil.no/rekkevidde used, this is practical guide, not officially calibrated.
4. This range is calculated using an EV range predictor from Norway used for the Nissan Leaf MK1 http://www.gronnbil.no/rekkevidde. This range predictor is a practical guide and not officially calibrated.

5. Although the Tesla model S has a much larger range (more than 400 km), which could reduce the number of charging points, the Tesla is in the highest price segment of the market and a significant number of vehicles of this type are not expected in Bhutan.
6. World Bank data 2012, Bhutan Transport Sector.

References

Barsukov, Y., and J. Qian. 2013. *Battery Power Management for Portable Devices*. Artech House.

Dinger, A., R. Martin, X. Mosquet, M. Rabl, D. Rizoulis, M. Russo, and G. Sticher. 2010. *Batteries for Electric Cars: Challenges, Opportunities, and Outlook to 2020*. Boston Consulting Group.

The Economist. 2014. *Tesla's Gigafactory, Driving Ahead*. March 3.

Egan, D.R., C. Ponce de León, R.J.K. Wood, R.L. Jones, K.R. Stokes, and F.C. Walsh. 2013. "Developments in Electrode Materials and Electrolytes for Aluminium–Air Batteries." *Journal of Power Sources* 236: 293–310.

Hübner, Y., P. Blythe, G. Hill, M. Neaimeh, J. Austin, L. Gray, C. Herron, and J. Wardle. 2013. *49,999 Electric Car Journeys and Counting*. Electric Vehicle Symposium and Exhibition (EVS27).

Kalmbach, R., W. Bernhart, P. Grosse Kleimann, and M. Hoffmann. 2011. *Automotive Landscape 2025: Opportunities and Challenges Ahead*. Roland Berger.

McDonald, P., J. Brady, M. O'Mahony, M. Sanmarti, M. Daly, S. McGrath, and N. Vierheilig. 2012. *Data Collection and Analysis in a Pan-European Electric Vehicle Fleet*. Washington, DC: 92nd Annual Conference of the Transportation Research Board, January.

Mock, P., and Z. Yang. 2014. *Driving Electrification: A Global Comparison of Fiscal Incentive Policy for Electric Vehicles*. Washington, DC: The International Council on Clean Transportation.

Moure, C., M.R. Roche, and M. Mammetti. 2013. *Range Estimator for Electric Vehicles*. Electric Vehicle Symposium and Exhibition (EVS27).

Rogers, E.M. 2003. *Diffusion of Innovations*. New York: Simon and Schuster.

Shirk, M., and J. Wishart. 2015. *Effects of Electric Vehicle Fast Charging on Battery Life and Vehicle Performance*. Idaho Falls: Idaho National Laboratory.

Siahrostami, S., V. Tripkovic, K.T. Lundgaard, K.E. Jensen, H.A. Hansen, J.S. Hummelshøj, J.S.G. Myrdal, T. Vegge, J.K. Nørskov, and J. Rossmeisl. 2013. "First Principles Investigation of Zinc-Anode Dissolution in Zinc–Air Batteries." *Phys. Chem. Chem. Phys.* 15: 6416–6421.

Trigg, T. and P. Telleen. 2013. *Global EV Outlook 2020*. IEA (International Energy Agency).

Yoo, H.D., E. Markevich, G. Salitra, D. Sharon, and D. Aurbach. 2014. "On the Challenge of Developing Advanced Technologies for Electrochemical Energy Storage and Conversion." *Materials Today* 17 (3).

CHAPTER 5

Fiscal and Economic Incentives

Key Messages

- International experience suggests that the effectiveness of fiscal incentives for electric vehicle (EV) uptake varies and that uptake depends on both price and nonprice factors.
- Current fiscal incentives in Bhutan are suitable for a low EV uptake target, with switching financially viable for a vehicle that drives more than 16,000 km per year.
- Future incentive programs to achieve more ambitious uptake targets will need balanced considerations on nonprice factors and increased fiscal support, along with sustainable EV market development in the long term.

International Experience with Incentive Programs

Various incentive programs to promote EVs have been adopted by national governments and subnational level authorities in countries around the world, including China, the United Kingdom, and the United States. Often national-level policies are put in place to lead the EV effort, mobilize resources, and set strategic direction, while state-level or city-level incentives complement the national policies by further facilitating consumers or providing additional incentives that are suitable to the local situation. The design of incentives is generally based on government policy objectives and specific market segments, such as for example research and development (R&D) subsidies for local car production or subsidies for high-mileage vehicles in city centers to improve local air quality (see also appendix B).

Incentives can be categorized as being oriented toward either the vehicle manufacturers or the consumers (table 5.1). Measures targeted at manufacturers are focused on financial support for R&D and vehicle development. For consumer financial incentives, most governments use a mix of cash and noncash incentives to promote EVs, including consumer purchase subsidies (cost subsidies) or measures such as taxes exemptions and tax credits. Other financial measures focus on creating differences in the cost of owning and driving an EV

Table 5.1 Financial and Nonfinancial Incentives

	Vehicle manufacturers	Consumers
Financial incentives	Grants, loans, and tax credits for R&D and vehicle development	Purchase subsidies Tax rebates Tax credits Exemption from taxes and charges Discounted tolls and parking fares
Nonfinancial incentives	Extra credit for calculating fuel economy for meeting national requirements	Preferential parking spaces Access to restricted roadways/highways Expedited permitting Installation of charging units

Source: EVI and IEA 2013.
Note: R&D = research and development.

compared to a regular car, for example by exempting the EV from charges normally applied to internal combustion engine (ICE) vehicles or providing discounts for toll and parking fees. London, for example, waives its congestion charges for EVs. Nonfinancial incentives for consumers can include preferred parking, access to certain driving lanes, free charging, and an expedited permitting process.

Internationally, the most popular financial incentives are direct cost subsidies and tax exemptions. Most European countries offer direct subsidies upon purchase of an EV, with amounts depending on the vehicle's emissions performance and battery capacity. The subsidies are often capped at a certain amount, and in some countries the number of vehicles receiving subsidies is also capped. To date, only Norway is offering a value added tax (VAT) exemption on EV purchases. Among countries, the value of the fiscal incentives varies between Nu 366,000 and Nu 600,000 (US$6,100 to US$10,000), representing about 20–40 percent of the EV retail price (figure 5.1).

As also shown in figures 5.1 and 3.1, variation exists in the effectiveness of fiscal incentives to create an EV market. Strong incentives are generally required to stimulate EV purchases, but these incentives will not necessarily result in a higher market share and growth rate if other complementary factors and measures are not in place. International experience suggests that some countries that put in place strong fiscal incentives do have higher market shares and growth rates, as observed in Norway and California (see figures 5.1 and 5.2), but in other countries, such as Denmark (figure 5.2), high incentives seem to have more limited success. It is also observed that more variation in performance exists among countries in the middle range, offering slightly lower incentives. Low customer EV awareness, customer preferences, and lack of charging infrastructure are possible explanations for the limited success in these cases. The effectiveness of fiscal incentives also depends on the taxes imposed on conventional ICE vehicles and the effective price difference between electric and ICE vehicles.

In addition to economic factors, consumer preferences also strongly influence purchase decisions, especially among early adopters. According to a McKinsey

Fiscal and Economic Incentives

Figure 5.1 Fiscal Incentives and Electric Vehicle Penetration Rate, Selected Countries

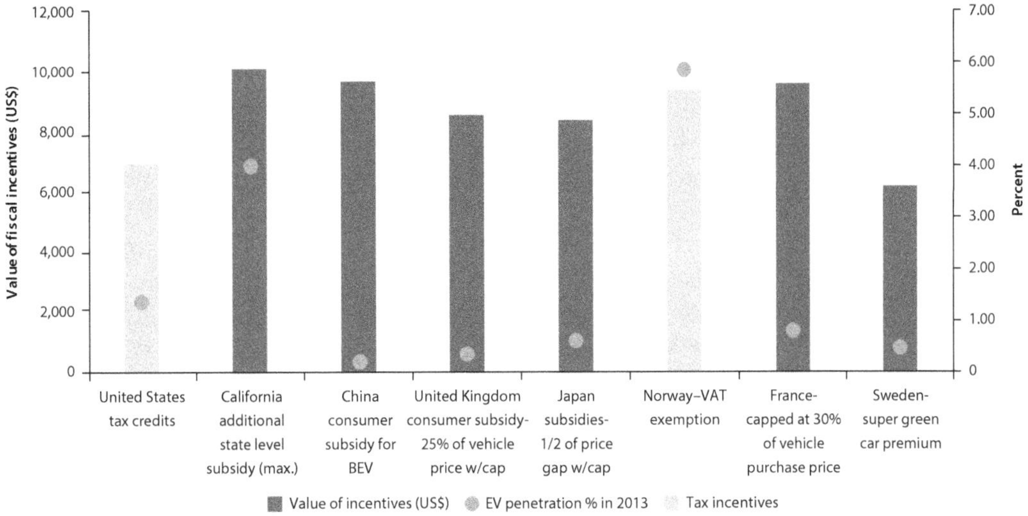

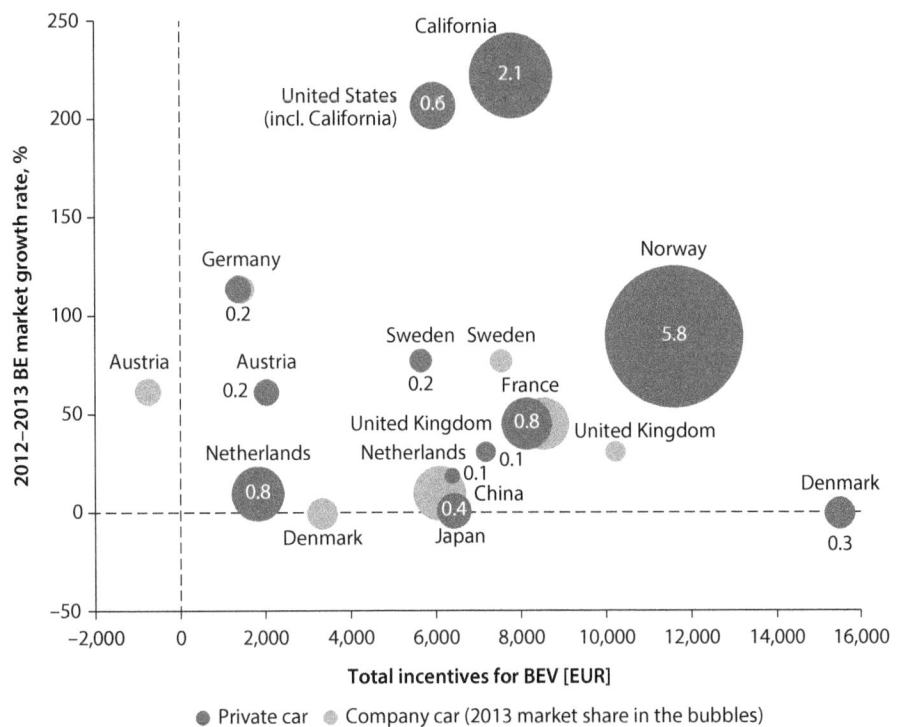

Source: EVI and IEA 2013; Institute of Transport Economics 2013; and ICCT 2014.
Note: Data show the maximum value of incentives offered in each country. BEV = battery electric vehicle; EV = electric vehicle; VAT = value added tax.

Figure 5.2 Market Growth Rate vs. Per-Vehicle Incentive for Renault Zoe Battery Electric Vehicle, Private and Company Cars, 2012–2013

Source: ICCT 2014.
Note: BEV = battery electric vehicle; EUR = Euros.

The Bhutan Electric Vehicle Initiative • http://dx.doi.org/10.1596/978-1-4648-0741-1

survey among early EV buyers in Japan (McKinsey 2012), the majority of EV buyers made the purchase because of the new green technology that is fulfilling their aspirations, with about half the buyers not concerned about fuel efficiency or available incentives. In contrast to the early adopters, people who were considering buying an EV but then decided to purchase a regular vehicle instead cited price, fuel efficiency, and design as the top reasons for not buying an EV. This group of potential buyers is more practical and puts more weight on price when making a purchase decision, exhibiting the attitudes of the mainstream car-buying public. The early adopters, estimated to comprise 1 percent of the car-buying population, are a higher income group, own more than two cars, and tend to drive only shorter distances. According to the same survey in Japan, a third of current EV users are less willing to purchase another EV; among the reasons cited were a lack of confidence in the dealer's capacity to service the vehicles, a lack of charging facilities for their apartments, and surprises related to the increase in electricity bills. In a study in Norway, the majority of buyers also said they bought an EV for environmental reasons (38 percent), economic reasons (29 percent), and practical reasons (28 percent) (Institute of Transport Economics 2013). In terms of disadvantages, in 2012 most EV owners in this survey were not satisfied with the range of their EV (figure 5.3). Experiences with the EV batteries, however, seem to be improving; in 2006 battery life and price were still cited by EV owners as the biggest problems.

In countries where EVs are promoted, heavy government subsidies are required. Public spending on fiscal incentives in 15 countries with EV initiatives amounted to US$3 billion (Nu 180 billion), constituting the second largest policy area for public support for EV after R&D (figure 5.4). In Norway, the public cost of the economic incentive program for consumers has so far amounted to US$141.4 million (Nu 8,466 billion) (figure 5.5). As economic incentives in Norway do not involve direct cost subsidies, the fiscal impact is characterized by loss in government revenues. In Norway and other developed economies, the government will be able to compensate for this loss in revenue by increasing rates for those who pay or by increasing taxes and charges in general to maintain the same revenue collection. This implies that the key impact of the incentives program in the case of Norway is not fiscal but rather distributive. This is more plausible in developed countries where the tax base is large enough for the government to exercise various fiscal measures without much concern for total revenue collection. In contrast, in developing economies like Bhutan, where the revenue base is narrower and less versatile, the government will face more limitations in exercising fiscal incentives because of significant impacts on state revenue. Box 5.1 presents an overview of Norway's fiscal and economic incentives for EV.

Analysis of Incentives and Total Cost of Ownership in Bhutan

Building on international experience, this section uses a total cost of ownership (TCO) analysis to study how the current fiscal incentives in Bhutan are working to decrease the TCO of EVs and make them more affordable, thus

Fiscal and Economic Incentives

Figure 5.3 Disadvantages of an Electric Vehicle Identified by Different Groups of EV Owners in Norway

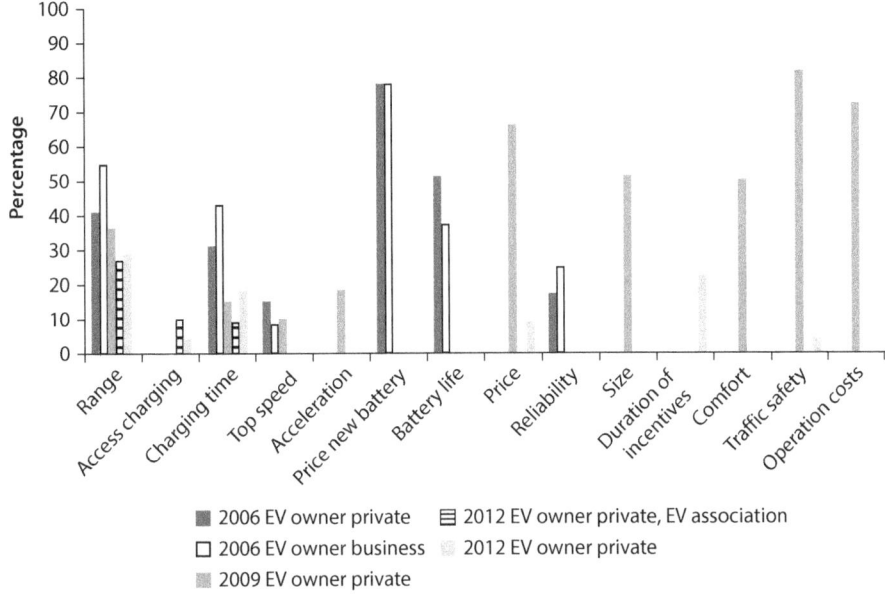

Source: Institute of Transport Economics 2013.
Note: EV = electric vehicle.

Figure 5.4 Combined Public Spending on Electric Vehicles in 15 Countries with EV Initiatives

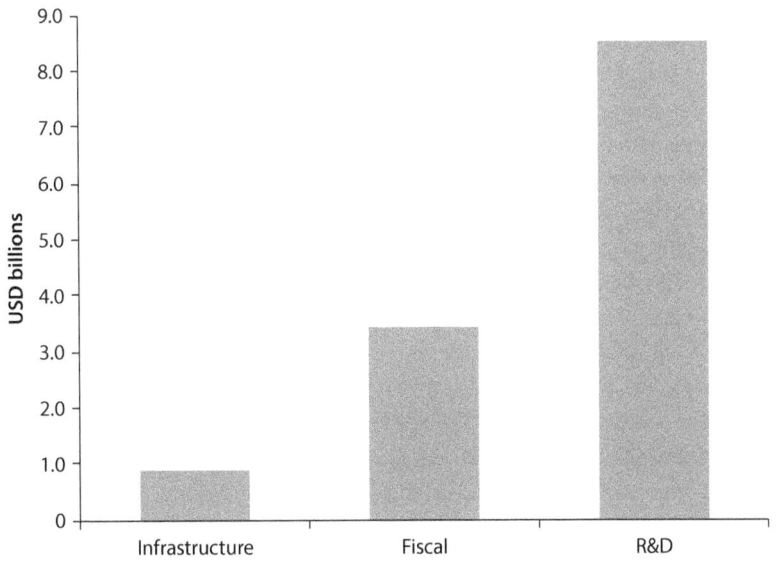

Source: EVI and IEA 2013.
Note: R&D = research and development.

The Bhutan Electric Vehicle Initiative • http://dx.doi.org/10.1596/978-1-4648-0741-1

Figure 5.5 Total Fiscal Support for Electric Vehicle Incentives in Norway

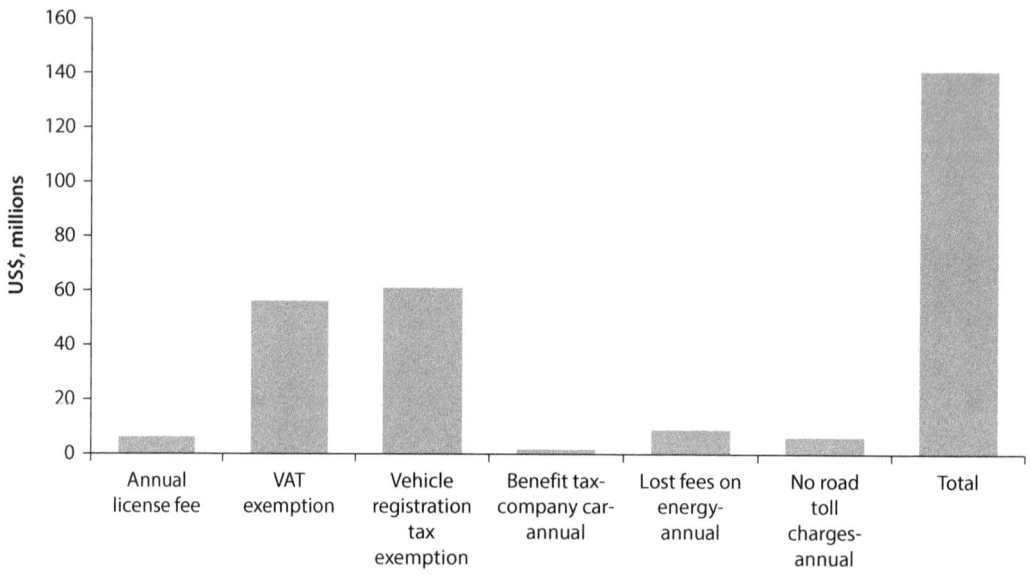

Source: Data from Institute of Transport Economics 2013 and compiled by the World Bank.

Box 5.1 Norway's Fiscal and Economic Incentive Programs

Norway is one of the leading countries in terms of EV market penetration. The country has adopted strong consumer incentive programs and was able to achieve a relatively high market share of EVs and market growth rates. Although various reasons exist for Norway's outstanding experience with EVs, this box focuses only on the design and effectiveness of its fiscal and economic incentives for EVs, to identify lessons that could be applicable in the context of Bhutan.

Table B5.1.1 presents an overview of the impact of Norway's economic incentive programs on the costs for electric and ICE vehicles. In addition to providing one of the highest fiscal incentive programs in the form of a VAT exemption to reduce the EV purchase price, the country also uses other incentives to further reduce both the upfront cost of purchasing an EV (for example by reducing vehicle registration) and operating costs throughout the vehicle's operating period. Operating costs, for example, are reduced by lowering fees for annual licensing, energy use, road tolls, and parking. Considering that these fees are typically very high in Norway, the total economic benefits for EV drivers are well above the upfront fiscal incentive.

Figure B5.1.1 illustrates the total reductions in upfront costs and annual operating costs for EV owners in Norway. It is estimated that their total benefit is about US$24,300 (Nu 1,458,000), with an upfront benefit of about US$19,500 (Nu 1,170,000) or 84 percent of the total benefits and an annual operating benefit of about US$3,700 (Nu 220,000) or 16 percent of total benefits. The most important incentive for reducing upfront costs in Norway is the VAT exemption, while the most important benefit for reducing operating costs is the exemption for Norway's high road toll charges.

box continues next page

Fiscal and Economic Incentives

Box 5.1 Norway's Fiscal and Economic Incentive Programs *(continued)*

Table B5.1.1 Economic Incentives for Electric Vehicles and Resulting Costs for EVs vs. Petrol Cars in Norway

Incentive measures	Cost for EVs	Cost for petrol cars
Annual license fee reduction	51 €	360 €
VAT exemption	6,875 € (approx. purchase price of 25,500 €)	3,750 € (approx. purchase price of 15,500 €)
Vehicle registration exemption	0 €	Small petrol car 2,500–3,100 € Compact car 5,600–9,400 €
Waiver of fees on energy use for EVs	Electricity cost without fees: 0.09 €/kWh (with fees 0.13 €/kWh)	Petrol fees 0.6 €/liter CO_2 charge on petrol 0.11 €/liter
No road toll charges	0 €	Approx. 1,000 € per year
Free parking in public parking spaces	0 € (savings unknown but could be significant)	Varies.

Source: Data from Institute of Transport Economics 2013 and compiled by the World Bank.
Note: € = euro; EV = electric vehicle; kWh = kilowatt-hour; VAT = value added tax.

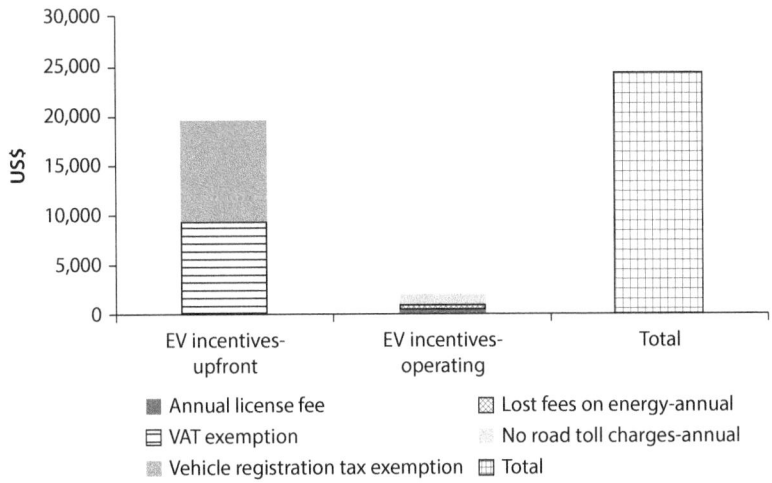

Figure B5.1.1 Economic Incentives for Electric Vehicle Owners in Norway

Source: Institute of Transport Economics 2013.
Note: EV = electric vehicle; VAT = value added tax.

It is important to note that Norway's successful EV initiative is not a result of its fiscal and economic incentives alone, and the country combines its strong financial incentives with free parking and free charging on the supply side. Other factors and consumer characteristics also influence purchase decisions, including vehicle technology, consumer driving needs, general EV awareness, consumer perceptions, and other specific local market factors. Although international experience suggests that costs, concerns about available charging infrastructure, and limited travel ranges are the top three factors discouraging the purchase of EVs, in Norway EV buyers have made their purchase decision

box continues next page

Box 5.1 Norway's Fiscal and Economic Incentive Programs *(continued)*

based on nonprice factors, such as green aspirations or fascination with a new technology. As second car purchases are prevalent in the country (42 percent of households own more than one car), EVs in Norway are typically purchased as a second car to complement—rather than fully replace—the household vehicle fleet. These other factors help explain the effectiveness of the incentives in Norway.

---

increasing their uptake. After a brief introduction to the methodology and key assumptions for the analysis in the context of Bhutan (describing, for example, the vehicles compared and factors such as incentives and the price of fuel and vehicles), the section presents the key findings of the TCO analysis using Bhutan's current incentives, describing the potential for vehicle buyers (private and taxi drivers) to switch to EVs. Next, the same analysis is used to see what additional financial incentives might be needed to achieve higher uptake targets. A final section describes the key policy considerations for designing an incentives program in Bhutan. Additional detail about the analysis and assumptions is provided in appendix C.

TCO Analysis Methodology and Key Assumptions for Bhutan

TCO refers to the cost of purchasing, operating, and maintaining a car throughout the vehicle's life. An analysis of TCO is useful to assess the financial viability of buyers switching to EV. Although other behavioral and nonprice factors also influence purchase decisions and impact the performance of incentives, the concept of TCO is widely used to specifically study the economic factors of a purchase decision and analyze the effectiveness of fiscal and economic incentives (ICCT 2014; World Bank 2011). Box 5.2 illustrates examples of TCO analyses used for several European countries.

For the TCO analysis for Bhutan, the current incentives are used as a baseline scenario to compare the TCO of an EV with that of a comparable ICE vehicle. To understand the switching potential for buyers, the basic assumption is that consumers will shift to EVs if the TCO of an EV is equal to or lower than that of a comparable ICE vehicle, with a greater likelihood of consumers making the switch when the difference is larger.

Key inputs and assumptions for the TCO analysis for Bhutan are as follows:

- **Composition of the EV fleet.** Different EVs were selected to analyze the TCO for the private vehicle market and taxi market.
  - *Private vehicle: Nissan Leaf at a discounted and full price.* This vehicle is available in the local EV market at a 30 percent subsidized price of Nu 1,260,000 (US$21,000) by the manufacturer. Information on prices for the marginal vehicle after the first lot is not available.
  - *Taxi vehicle: Refurbished Nissan Leaf and regular Nissan Leaf.* Among the locally available models (brand new Nissan Leaf, refurbished Nissan Leaf,

and Mahindra Reva e2o), the most suitable model was the refurbished Nissan Leaf (driven for about 20,000–30,000 km)[1] of which the price at around Nu 800,000 (US$13,333) is affordable for taxi drivers and that also has suitable capacity for taxi operations. Given that these vehicles may be available in limited quantities and given the environmental concerns of importing a secondhand vehicle, the TCO calculation was made for both the refurbished Nissan Leaf with its reduced price and the regular Nissan Leaf.

- **Choice for comparable ICE vehicle.** For the private vehicle segment, a Hyundai i20 was selected as a comparable ICE vehicle; the vehicle's product dimensions are relatively close to those of the Nissan Leaf, and the car is available in the local market. For the taxi segment, the Maruti Alto, which currently dominates the taxi fleet, was selected as the comparable ICE vehicle.
- **Current fiscal incentive programs.** The current fiscal incentives program, which is used to calculate the baseline scenario for the TCO analysis, consists of an exemption of the sales tax and customs duty; EVs are also not subject to the green tax. Taxes on conventional fuel vehicles are thus key policy variables that will influence the cost comparison between EVs and ICEs. The vehicle import ban was lifted in July 2014. The new effective tax rates for vehicles below 1,500 cc as of July 2014 are shown in table 5.2.
- **Fuel cost.** The analysis used a fuel cost of Nu 63 (US$1.05) per liter and an electricity cost of Nu 2.46 (US$0.04) per kWh.[2] In addition, a sensitivity analysis was carried out to understand the impact of fuel price increases of 1 and 7 percent.
- **Vehicle financing.** It is assumed that an average consumer will finance a vehicle through a loan up to 50 percent and personal savings up to 50 percent at the weighted average cost of capital of 10 percent (see also appendix C).
- **Discount rate.** A discount rate of 10 percent was selected for the TCO analysis. The 10 percent discount rate reflects the weighted average cost of capital of a vehicle purchase for an average consumer (see vehicle financing just above).

Box 5.2 TCO and Incentives in Other Countries

An analysis of electric and ICE vehicles' TCO and a comparison among countries can illustrate the variety of options and impacts (in terms of price reduction or market uptake) of incentive measures. The three figures in this box all compare the TCO for an EV (the Mitsubishi j-MiEV or the Renault Zoe) with that of a comparable ICE vehicle (the Fiat 500 or the Renault Clio) in various European countries.

Figure B5.2.1, which compares the TCO for the electric Mitsubishi i-MiEV with that of a comparable Fiat 500 1.2 gasoline car in Denmark, Norway, and Sweden, shows that only in Norway is the TCO for this EV below that of the Fiat. The difference of about 20–25 percent is mainly due to the road toll and parking fees incentive. Figure B5.2.2 shows a similar finding for the

box continues next page

Box 5.2 TCO and Incentives in Other Countries *(continued)*

Figure B5.2.1 TCO for Mitsubishi i-MiEV (EV) vs. Fiat 500 (ICE), 2012 Pricing, Comparing Norway, Sweden, and Denmark

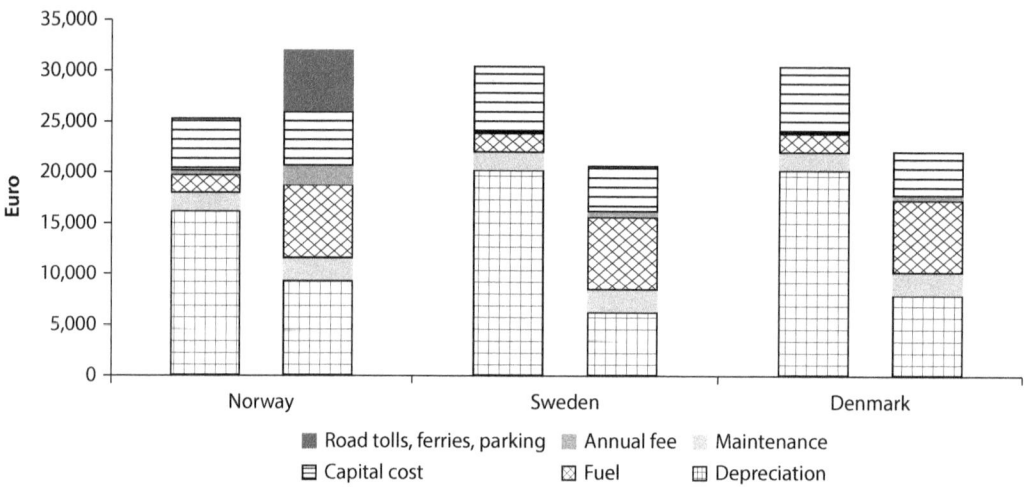

Source: Hannisdahl, Malvik, and Wensaas 2013.
Note: The calculation is for five years of ownership and 15,000 km per year. Estimates only.

Figure B5.2.2 Evaluation of TCO for France, Germany, and Norway

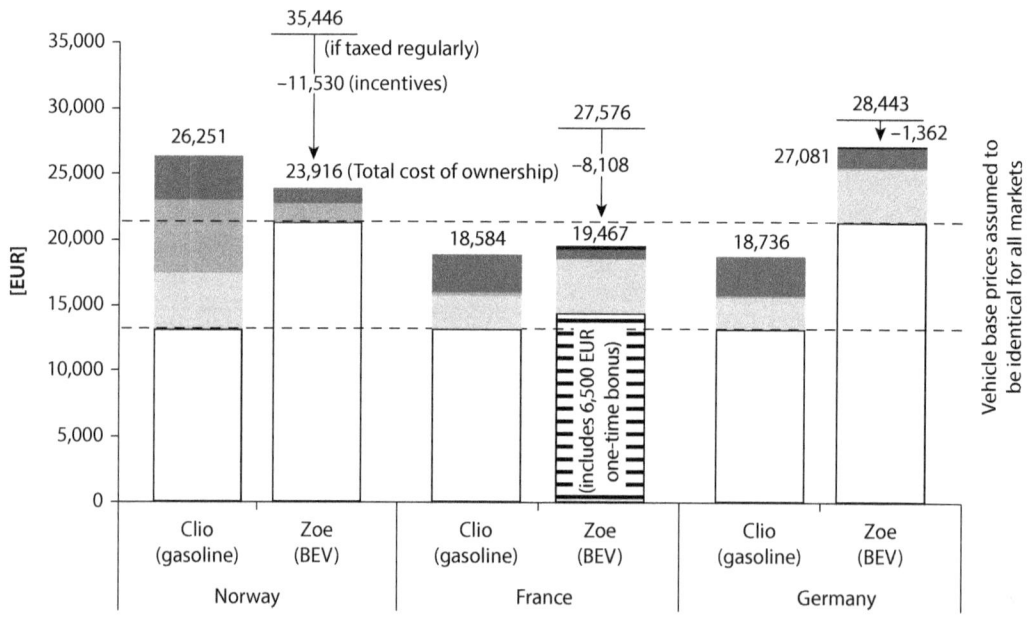

Total cost of ownership includes vehicle purchase and registration costs, as well as ownership taxes and fuel/electricity costs for 4 years. All data estimates for tax year 2013.

Source: ICCT 2014.
Note: BEV = battery electric vehicle.

box continues next page

Fiscal and Economic Incentives

Box 5.2 TCO and Incentives in Other Countries (continued)

Figure B5.2.3 Summary of TCO Calculations for Renault Clio vs. Renault Zoe (Private Car Market)

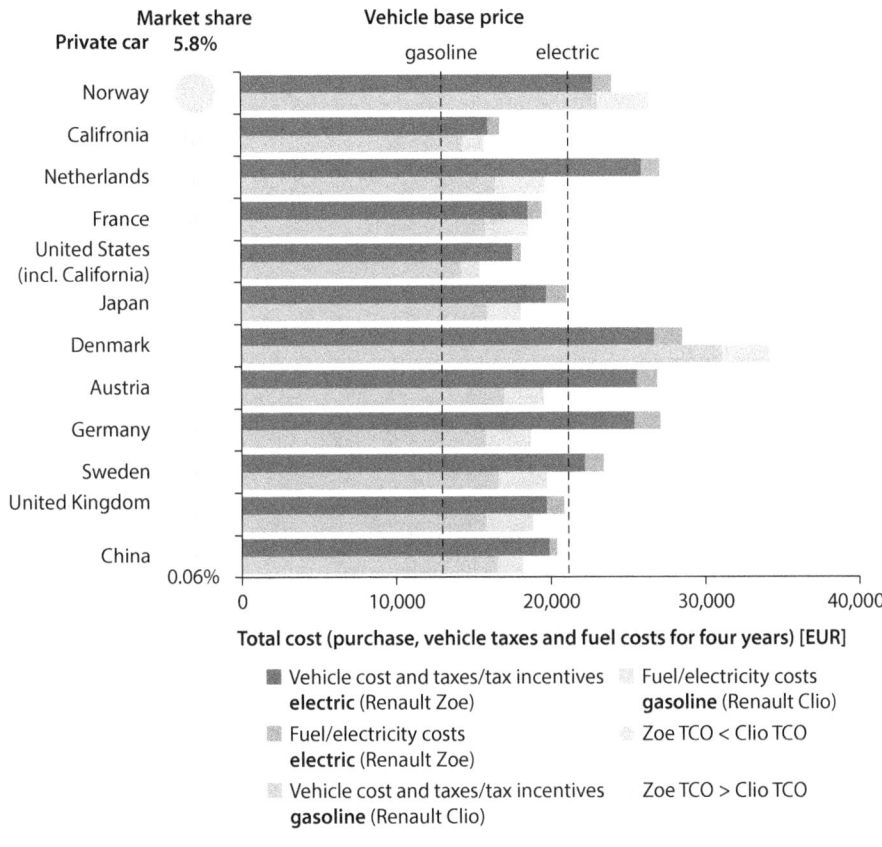

Source: ICCT 2014.
Note: TCO = total cost of ownership.

Renault Clio in Norway when compared to France and Germany. Again only in Norway is the TCO for the EV below that of the ICE vehicle, by about 9 percent.

Figure B5.2.3, in addition to comparing the TCO for the Renault Clio and Zoe, also illustrates the overall EV market share for a broader set of countries. The figure underscores that the TCO comparison between the electric and ICE vehicles is not the only factor determining the market share. Among the various high EV uptake markets (for example, the Netherlands, Norway, and California), only Norway actually has a lower TCO for EVs. In contrast, Denmark, which is the only other country with a lower TCO for EVs, has only a moderate EV uptake rate.

Table 5.2 Current Fiscal Incentives for Electric Vehicles—Taxes on Selected Vehicles Used in the Analysis

	Taxes applicable to ICE as of July 2014	Tax exemptions for EV
Sales tax (%)	45%	45%
Green tax (%)*	10%	EV not subject to Green Tax
Customs duty (%)	45% (except for vehicles imported from India)	45% (except for vehicles imported from India)
Total taxes (%)	55% for vehicles imported from India	45% tax exemption for India imported vehicles
	100% for vehicles imported from countries other than India	90% tax exemption for vehicles imported from countries other than India

Note: Taxes are applicable to the Hyundai i20, which is used as the ICE vehicle in the TCO analysis.
EV = electric vehicle; ICE = internal combustion engine, TCO = total cost of ownership.

- **Battery replacement.** Battery replacement will largely depend on the battery usage of an individual vehicle and mileage driven. For private vehicles, given their average annual mileage, it is assumed that no battery replacement is needed during the eight-year life of a vehicle. For taxis, given their high annual mileage, battery replacement is likely over this same operating period. The TCO analysis uses two key assumptions for taxi battery replacement:
    - *High annual mileage:* If the annual mileage of taxis is 50,000 km and above, battery replacement will be required in the third year and will cost Nu 165,000 (US$2,750).
    - *Low annual mileage:* If annual mileage of taxis is less than 20,000 km, battery replacement will not be required during the eight-year operating period.

Additional information about the assumptions and TCO calculations is provided in appendix C.

TCO Analysis for Private Vehicles and Taxis Using Current Incentives

Using the current incentives, key findings from the TCO calculation suggest that under current conditions the key variables for potential savings using EV are the vehicle's annual mileage and the upfront cost of a comparable ICE model (or alternatively, the cost of an EV). An economic case to switch to EV will exist when the TCO of a comparable ICE is higher than the TCO of an EV or—in other words—when the net savings (the difference between the TCO of the comparable ICE and an EV) is positive.

Private vehicles. For private vehicles and using the 30 percent discounted price for the Nissan Leaf, the break-even point (when the TCO of the EV is the same as the TCO of the comparable ICE vehicle) under the current tax incentives is when the vehicle's annual mileage is above 15,000 km per year and the price of the comparable ICE vehicle (after tax) is above Nu 945,500 (US$15,758). This average mileage is above the current estimate of 10,000 km per year as the average mileage for private vehicles in Thimphu. When using the full price for a Nissan Leaf—considering that the subsidized price may be temporary—results show that the 30 percent increase in the price of the EV would lead to a significant increase in TCO. Consequently, a much higher mileage (over 33,000 km) would be needed to achieve positive savings. Similarly, the price of the comparable ICE vehicle would need to be much higher (around Nu 1,457,000 or US$24,283) to make the EV affordable.

Because the savings by private vehicles are sensitive to assumptions on fuel price and discount rate, a sensitivity analysis was conducted to examine the impact of changes in either one of these assumptions. For the increase in fuel price, the sensitivity analysis considered both a 1 percent increase (reflecting the global oil price projection for 2025; see appendix C) and a 7 percent increase (an increase more in line with historical increases in Bhutan). As shown in figures 5.6 and 5.7, at a discount rate of 10 percent (similar to the previous analysis), the annual mileage needed to create positive savings when assuming a 7 percent annual fuel price increase is around 14,000 km (figure 5.6), while this increases to over 16,000 km if the annual fuel price increase is assumed to be only 1 percent (figure 5.7). Discount rates also have a significant impact on the results. With an assumption of a 7 percent annual fuel price increase and a 5 percent discount rate, the required mileage to achieve savings is over 11,000 km, while at a 15 percent discount rate but the same fuel price increase, required mileage reaches 17,000 km.

Taxis. In the case of taxis, the analysis looked at several options, including both the refurbished Nissan Leaf and a full price Nissan Leaf, as well as the implications of the taxis' driving patterns in terms of battery replacement

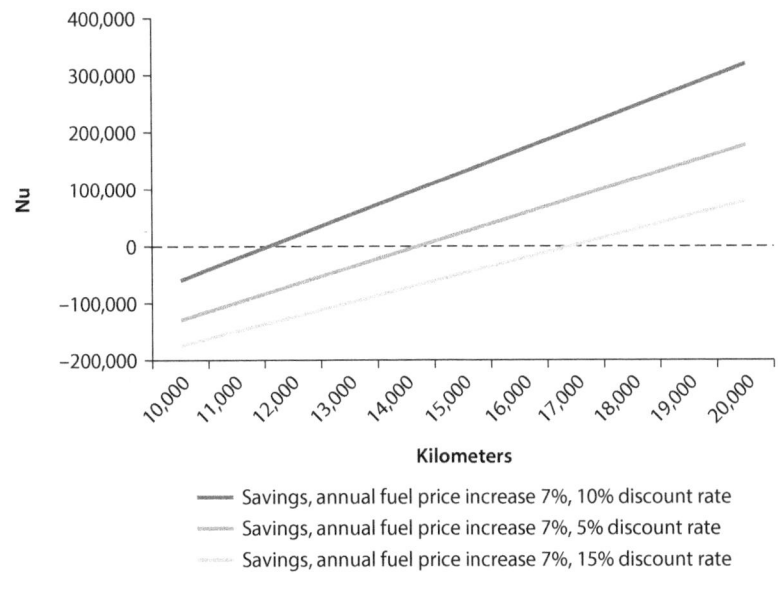

Figure 5.6 Savings for Private Vehicles when Switching to EVs, Annual Fuel Price Increase of 7 Percent at Different Discount Rates

Source: World Bank analysis.
Note: EV = electric vehicle; Nu = Bhutanese ngultrum.

Figure 5.7 Savings for Private Vehicles when Switching to EV, Annual Fuel Price Increase of 1 Percent at Different Discount Rates

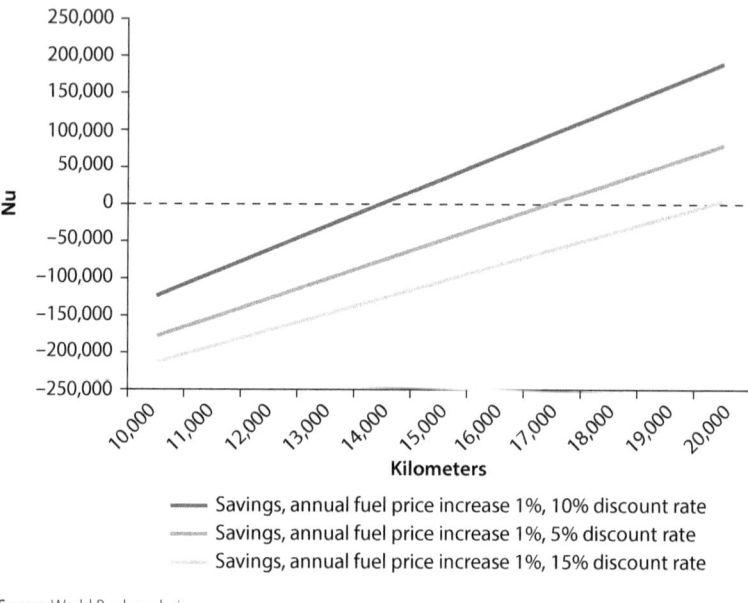

Source: World Bank analysis.
Note: EV = electric vehicle; Nu = Bhutanese ngultrum.

needs (that is, 50,000 km with one battery replacement and 20,000 km with no battery replacement).

Overall, the key findings from the TCO analysis suggest that, with the current incentives and because of the high annual mileage of taxis of about 50,000 km, savings are positive and a strong financial case exists for taxis to switch to EVs. Results, however, suggest that an affordable and suitable EV model for the taxi fleet will be critical because the difference in results between the refurbished and full-price Nissan Leaf are large (see figures 5.8 and 5.9). Moreover, mileage is a key factor: if a taxi drives 50,000 km per year, the savings will be large enough to cover the battery replacement cost. In contrast, if a taxi drives a much lower mileage of 20,000 km per year, the savings will be negative even without a battery replacement. Similar to the findings for the private vehicles, any savings are sensitive to assumptions on future fuel price increases and the discount rate. With an assumed annual fuel price increase of 7 percent, savings for taxi drivers will be positive with all three discount rates except in the case where a full-price Nissan Leaf drives only 20,000 km per year. However, if the assumption on annual fuel price increase is reduced to 1 percent, the savings will also be negative in the case of the full-price Nissan Leaf driving 50,000 km at 10 percent and 15 percent discount rates.

The TCO calculation does not take into consideration any revenue impact for taxi drivers from switching to EVs, for example as a result of time spent charging the EV. Other key challenges in the taxi segment are the availability of suitable

Fiscal and Economic Incentives

Figure 5.8 Savings for Taxis using Various EV Options, Annual Fuel Price Increase of 7 Percent at Different Discount Rates

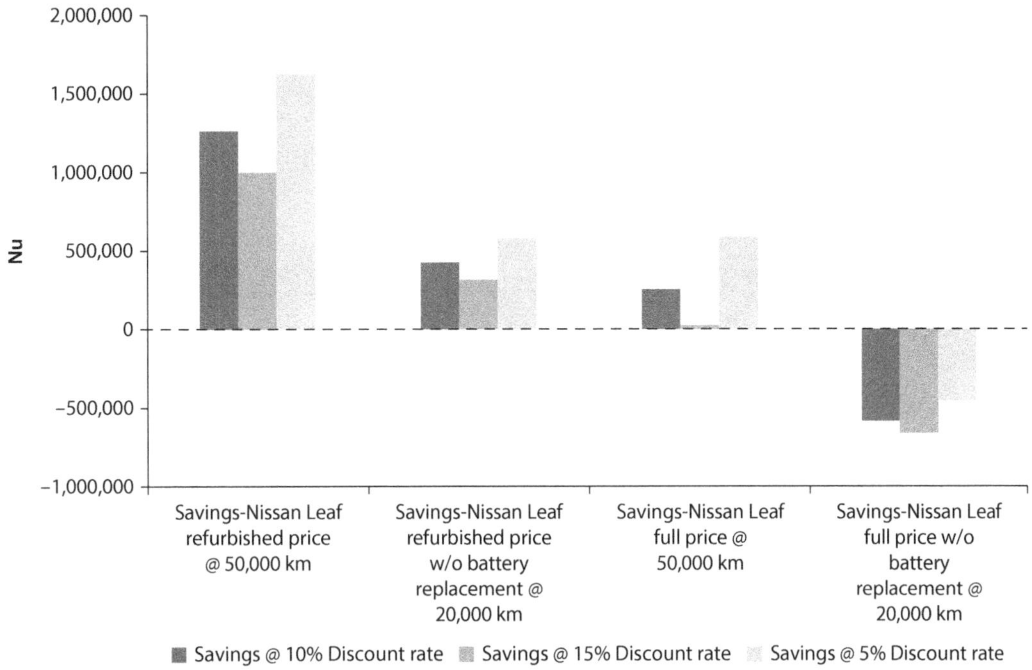

Source: World Bank analysis.
Note: EV = electric vehicle; Nu = Bhutanese ngultrum.

vehicles and specific user characteristics and requirements, such as the driver's ability to pay for higher upfront costs, access to vehicle financing, and parking availability in public housing areas.

TCO Analyses and EV Uptake Scenarios

The TCO analysis to study the impact of current incentives on the TCO of an EV and a comparable ICE vehicle can also be used to study the impact of potential future levels of fiscal incentives. Assuming more vehicle buyers will switch to buying an EV when the TCO of the EV becomes comparable or even drops below the price of the comparable ICE vehicle (see table 5.3), the level of subsidy required to achieve a certain uptake level can be calculated (see appendix C).

Although, based on the calculations, the current fiscal incentives are expected to be adequate to achieve the low uptake scenario with a target of 323 EVs for private use, achieving the higher EV uptake rates (969 and 1,615 vehicles in the high and super high uptake scenarios, respectively) would require additional incentives to attract more price-sensitive consumers. As tax exemptions are already used as a fiscal incentive, the analysis assumed that an additional incentive would be in the form of a cost subsidy on top of the existing tax exemptions. The results indicate that the level of cost subsidy required to meet the EV uptake

Figure 5.9 Savings for Taxis using Various EV Options, Annual Fuel Price Increase of 1 Percent at Different Discount Rates

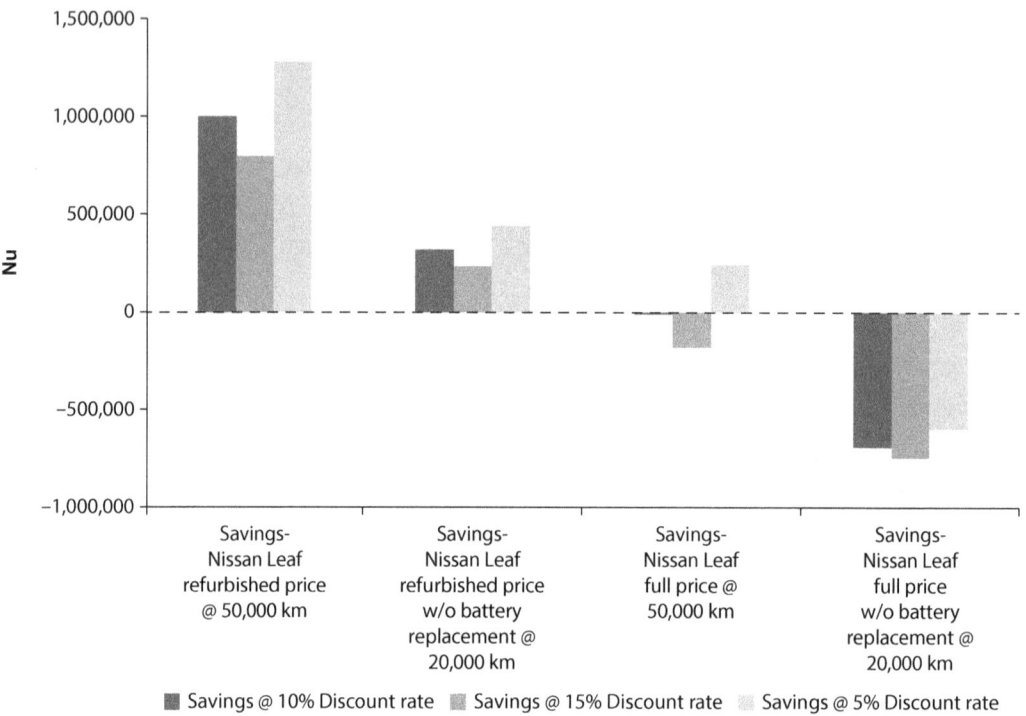

Source: World Bank analysis.
Note: EV = electric vehicle; Nu = Bhutanese ngultrum.

Table 5.3 Required Levels of Additional Cost Subsidy to Achieve High and Super High Uptake Scenarios

	Low uptake scenario	High uptake scenario	Super high uptake scenario
TCO	People buy EV with higher TCO $TCO_{EV} > TCO_{ICE}$	Incentives bring down TCO of EV to a comparable level or less than ICE $TCO_{EV} \leq TCO_{ICE}$ (±5%)	Incentives bring down TCO of EV to much lower level than TCO of ICE $TCO_{EV} < TCO_{ICE}$ (−20 to −30%)
Target number of EVs for private vehicles	323	969	1,615
Level of required cost subsidy (percentage of vehicle price)	0%	10%	35%

Source: World Bank analysis.
Note: EV = electric vehicle; ICE = internal combustion engine; TCO = total cost of ownership.

target would be 10 percent of the vehicle price for the high uptake and 35 percent for the super high uptake scenario.

Policy Considerations

According to the analysis, the incentives currently in place in Bhutan are suitable for a low uptake scenario in which a limited number of consumers will decide to buy an EV even as the TCO of the EV is higher than the TCO of the comparable ICE vehicle. With the current incentives, vehicles in the private vehicle market with an average annual mileage above 15,000 km have a strong economic incentive to switch to EVs. For the taxi market, a strong financial case exists for vehicles with an annual mileage of 50,000 km, although especially in the case of taxis other factors will be important given the unique characteristics of taxi operations compared to private vehicles, such as the need for fast charging during the day, access to vehicle financing, the need for parking availability in public housing areas, revenue impact, and marketing opportunities. More information will be needed to develop an appropriate incentive and strategy program for taxis.

As described in this section, the current level of incentives in Bhutan in terms of the percentage of vehicle price (100 percent) is high when compared to incentives in other countries. The large price gap between ICE vehicles and EVs despite this high level of incentives results from the currently small EV market and the limited local availability of suitable models. If more affordable EV models are available in the local market, consumers will have a better choice and more opportunities to switch to EV without additional fiscal incentives.

With the current incentive program in place, policies at this stage should focus more on nonprice factors, in particular on building consumer awareness and improving communications to provide reliable information to consumers about EVs. In addition, policies should focus on investing in charging infrastructure to support the development of a sustainable EV program. To achieve more ambitious targets, a future incentive program needs balanced considerations on nonprice factors, fiscal support, and sustainable EV market development in the long term. Consumer preferences and their response to price will determine the effectiveness of fiscal incentives. To design new incentive strategies, a better understanding of local consumer preferences will be needed, and both price and nonprice factors should be analyzed and included in the strategy.

Finally, other important policy considerations when designing an incentive program are the fiscal capacity of the government and achieving market development. The TCO analysis used targets for each uptake scenario, but these targets may not match the fiscal capacity of the government given competing priorities for public funds. Fiscal capacity to carry out incentive programs is a key differentiating factor between high-income countries and low-income countries. The government may wish to set some cap on the level of public support or funding available for the EV program, which can then be used as the key parameter when developing the targets (either as fixed amount or as percentage of vehicle price and with a cap on the number of vehicles).

To ensure sustainability, incentives should be designed with a long-term view of EV market development. Caution is needed when using too-heavy incentives in the early period of market development. If the price is too heavily subsidized, the market may take off quickly, but there may be adverse impacts on the EV market development in the longer run (besides the fiscal burden) as subsidies and incentives have to phase out sooner or later. If the price difference between the incentive period and the post-incentive period is too high, the market may not be able to sustain itself without the incentive. If the government wishes to develop the EV market in the long run, it may be better to take a gradual approach to establish the market and keep the incentives at a level that will not disrupt the market in the post-incentive period.

Notes

1. On March 10, 2014, as per letter C-3/26/325, the Royal Government of Bhutan (RGoB) approved the import of secondhand electric Nissan Leaf vehicles with mileage less than 30,000 km to be used as taxis in the country. The letter further states that it partially modifies the standing ban on import of all secondhand automobiles regardless of origin issued in COM/02/99/48 dated December 6, 1999. The order further states that the government needs to buy the existing taxis and sell them out of the country through an arrangement.
2. In July 2014, the government also started to impose a 5 percent tax on fuel in an attempt to curb fuel growth and align Bhutan's fuel prices with those in India.

References

EVI and IEA (Electric Vehicle Initiative and International Energy Agency). 2013. *Global EV Outlook: Understanding the Electric Vehicle Landscape in 2020*. Paris, France: EVI and IEA.

Hannisdahl, O.H., H.V. Malvik, and G.B. Wensaas. 2013. "The Future Is Electric! The EV Revolution in Norway—Explanations and Lessons Learned." Paper prepared for the Electric Vehicle Symposium and Exhibition (EVS27), Barcelona, Spain, November 17–20.

ICCT (International Council on Clean Transportation). 2014. *Driving Electrification: A Global Comparison of Fiscal Incentive Policy for Electric Vehicles*. Washington, DC: ICCT.

Institute of Transport Economics. 2013. *Electromobility in Norway: Experiences and Opportunities with Electric Vehicles*. Oslo: Institute of Transport Economics Norwegian Centre for Transport Research.

McKinsey. 2012. *Profiling Japan's Early EV Adopters*. McKinsey & Company.

World Bank. 2011. *The China New Energy Vehicles Program: Challenges and Opportunities*. Washington, DC: World Bank.

CHAPTER 6

Charging Infrastructure and Network Planning

Key Messages

- Different charging standards and plugs currently exist on the market. Because of the relatively small size of the electric vehicles (EV) market in Bhutan and strong economic ties with India, aligning with standards in India (expected in 2016) may be beneficial.
- Worldwide, most charging of EVs is done at home and at work. In Bhutan, to reduce costs and limit the number of stakeholders involved, initial focus should be on stimulating home and workplace charging. A healthy business case for public charging is not yet available, and defining roles for the various stakeholders that would be involved can take a long time.
- Fast chargers add additional value to the normal charging network. A phased rollout targeting specific user groups is necessary to develop a business case.
- In the low EV uptake scenario, by 2020 about 648 normal chargers (for home, work, and public charging) and about 10 fast chargers will be required. In the super high uptake scenario, these number increase to 7,000 normal and 240 fast chargers. Estimated infrastructure costs for these scenarios are about US$684,000 (Nu 38,880,000) for the low and US$11,131,000 (Nu 667,860,000) for the super high EV uptake scenario.
- A hybrid market model (through a public-private partnership, or PPP) should be created for a dedicated charging infrastructure operator.
- To realize the needed infrastructure for public charging, a business model for charging must be developed with an active role or incentive program by the Royal Government of Bhutan (RGoB).
- The current grid is likely sufficient for charging needs under the low EV uptake scenario (with 0.4 percent of the total peak from EV charging), but the grid does need constant development and further investigation is necessary. In the super high EV uptake scenario, an estimated 5.2 percent of the total peak will stem from EV charging, which means the impact of wider EV uptake will inevitably be noticeable and off-peak EV charging will be necessary.

The Bhutan Electric Vehicle Initiative • http://dx.doi.org/10.1596/978-1-4648-0741-1

Importance of Charging Infrastructure

The need for regular charging is an integral part of the introduction of EVs. International experience, however, suggests that it is one of the toughest technical, financial, and organizational challenges to overcome. For internal combustion engine (ICE) vehicles, fueling infrastructure is already in place and oil companies have a solid business model. For EVs, however, the charging infrastructure is new and needs to be built from scratch, even while using the existing electricity grid.

The importance of having charging stations is illustrated by figure 6.1, which shows a direct and positive relationship between the availability of charging stations and the uptake of EVs. Despite the overall positive correlation, however, several discrepancies exist in the data, such as for example in Ireland where the EV market share is much lower than expected considering the number of charging stations per 100,000 residents. In this case, the local Distribution System Operator (DSO) has very actively developed the number of charging stations to prepare the national electricity grid for the expected uptake of EVs in coming years. Other discrepancies between the size of the market share and the number of charging stations might also be explained by (a) the location and subsequent (poor) use of the charging infrastructure[1] and (b) a possible lack of inclusion of semipublic and private charging points in the data. The data source does not provide specific information about these charging points and most countries do not have accurate data about the numbers of semipublic and private charging.

Figure 6.1 Correlation between Number of Charging Stations and EV Uptake per Country

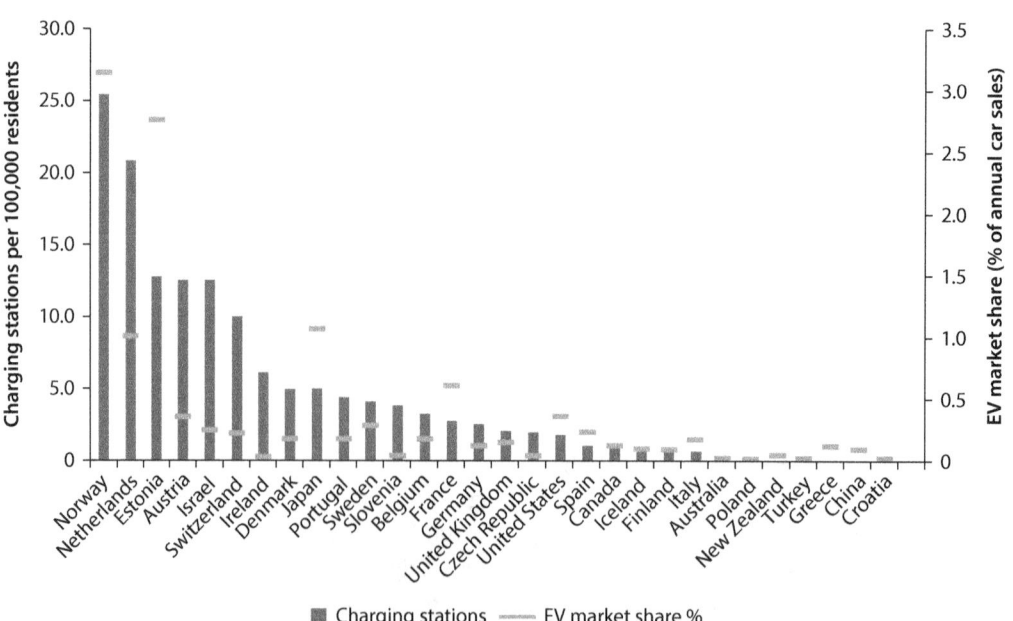

Source: Sierzchula et al. 2014.
Note: EV = electric vehicle.

Charging Infrastructure and Network Planning

For a successful introduction of EVs in Bhutan it is essential that charging infrastructure is in place. The various components involved (infrastructure planning and investment, operation and maintenance of facilities, and adjustments to the electricity system) will all require government attention along with private investments.

Types of EV Charging and Available Standards

Currently, different charging standards and types of EV charging are available. In terms of the technical aspects, a distinction can be made between normal and fast charging, each with its own infrastructure needs and use of different currents (alternating current [AC] or direct current [DC]). In terms of location and use, a distinction is made between home, semipublic (including workplace), public, and on-route fast charging. In addition, alternative charging options are available. Globally, home and workplace charging are the most commonly used types of charging.

Charging Types Based on Location

Based on where an EV is charged, a distinction can be made between home, semipublic, public, and on-route fast charging. Table 6.1 presents an overview of these charging types. Workplace charging typically takes place at a semipublic or private location, while on-route fast charging along highways would typically be public. As will be discussed in "EV Charging Options in Bhutan," below, which examines these charging options in the context of Bhutan, home charging on private ground is not an option for all EV users in Bhutan, in which case other solutions must be provided.

Among the four options, charging at home (a private location) and work (often semipublic/private locations) are widely seen as the standards for EV charging

Table 6.1 Types of EV Charging Based on Location and Type

Home/private	Semipublic	Public	On-route fast charging
• Charging on private ground. • Overnight and during weekends (minimum of 4–8 hours). • Charging station or use of existing standard 16A socket. • Requires certified installation or check to ensure safety.	• Charging at public locations on private ground (e.g., retail locations, office buildings, parking at apartment blocks). • Includes some workplace charging. (Office buildings may have discount contract with energy provider). • Retail stores may offer free charging as a service to customers.	• Services for visitors or EV drivers without private charging options. • Requires dedicated parking places at a public charging station. • Increases EV visibility. • Involves various stakeholders. • Charging station requires robust outdoor setup with payment system and separate grid connection. • High installation and operation costs.	• Enables longer travel distances. • Setup similar to existing networks of gas stations, with charging stations at intervals of 50–75 km in and around cities with many EVs. (Traffic density can be used as pointer.) • Typically higher speed of charging than at other locations.

worldwide. With vehicles typically used for average commuter trips of 20–30 km, home and workplace charging are sufficient. Charging when the vehicle is parked also has less impact on the normal usage pattern. On-route fast charging can enable a longer journey, but currently is only a small part of the total charging transactions worldwide. In general, charging locations (in particular semi-public and public stations) must be clearly marked and dedicated and accessible to EVs only. Reliable options for charging are important for the acceptance and uptake of EVs.

Charging Type and Speed

Charging an EV takes more time than refueling an ICE vehicle and currently still involves a variety of standards. In general, charging is done by connecting the EV to a charging point (normally the regular electricity grid) using a charging cable. The time for a full recharge depends on factors such as the grid connection and the type of EV, charging station, socket, and charging cable. As shown in table 6.2, three types of charging can be distinguished in terms of speed, namely normal, semifast, and fast charging. Not all charging types are available for every vehicle. The Nissan Leaf, for example, comes standard with 3.7 kW (kilowatt) AC and 50 kW DC charging possibilities.

The charging standards are the same for full electric vehicles (FEVs) and plug-in hybrid electric vehicles (PHEV). DC fast charging, however, is not a logical choice for PHEVs because they can use the ICE for range extension.

In general, safety is an important aspect of the charging process. For all types, the product and its installation and maintenance should be monitored to ensure safe operations. This especially applies to home charging where the electricity connection is made at the household level. A dedicated charging point (socket) is recommended to prevent grid overload.

Standardization

For both normal charging (using AC) and fast charging (using DC), different standards have been developed and are in use in the United States (type 1, CCS [Combined Charging Standard], and Tesla), Europe (type 2 and CCS), Japan (type 1 and CHAdeMO), and China (GB/T). India, whose decision has potentially large implications for EVs in Bhutan, has not yet formally decided on the charging standard for EVs.

Table 6.2 Types of Charging and Charging Speeds

Type of charging	Location	Current	Standards	Power (kW)	Time to full battery	Indicative costs of charging infrastructure (US$)
Normal	Home, semipublic, and public	AC	Household, Universal, Type 1, Type 2-mode 3	2–4	3–6 hours	50–1,500
Semifast	Semipublic	AC	Type 2-mode 3 and Combo	11–22	0.5–1 hours	1,500–5,000
Fast	On route	AC/DC	Combo, CHAdeMO, Tesla, and GB/T	40–50	15–30 min.	10,000–20,000

Source: World Bank analysis.
Note: AC = alternating current; DC = direct current.

Charging Infrastructure and Network Planning

Table 6.3 summarizes current charging standards for normal and DC fast charging, listing the most common types and modes. Mostly a distinction is made between the "types" of plug and socket in combination with the charging "mode." The charging modes differentiate themselves in relation to the sockets used, maximum charging speeds, and communication possibilities (table 6.4). Standards for the vehicle side of the charging cable will be determined by the origin of the vehicle. Photo 6.1 illustrates examples of charging stations. Appendix D provides more information about the suppliers of CHAdeMO fast chargers. Box 6.1 provides a first assessment on charging standards in Bhutan.

Alternative Charging Types

Although conductive charging is the worldwide standard, the technology is still in an early phase and will develop more over time. Two other ways to charge an EV are battery swapping and induction.

Battery swapping. Battery swapping eliminates the disadvantage of charging time by changing the empty battery for a full one. However, the high investment cost for swapping stations, a lack of standardization of battery packs, and intellectual property rights by original equipment manufacturers

Table 6.3 International Standards for Plugs and Sockets for Normal and Fast Charging

Normal charging	Fast charging
General: • Charging cable is normally supplied with the vehicle and carried in the car. • Charging standards are described in IEC 62196, the international standard for set of electrical connectors and charging modes for electric vehicles.	**General:** • Fast charging stations have a fixed cable because of safety requirements considering the high voltage. • Globally, different standards have developed.
Sockets and plugs: • **"Normal" household socket** (not recommended for safety reasons). Can be used for almost every EV. No communication between the vehicle and the charging point. Not suitable for smart grid applications. • **Industrial plug (IEC 60309).** Designed for industrial use and weatherproof plug for 230 V. No communication between the vehicle and the charging point. • **Type 1 "yazaki" socket.** Japanese and American standard for normal charging. Communication between the vehicle and charging point. Can charge up to 7.4 kW (32A, 1-phase). (the United States uses separate standard JSAE 1772 because of the different voltage standards.) Suitable for smart grid applications. • **Type 2 "mennekes" socket.** European standard for normal and semi-fast charging produced in Germany. Communication between the vehicle and charging point. Can charge up to 44 kW (63A, 3-phase). Suitable for smart grid applications. • **Type 3 "le grand" socket.** French/Italian standard. Can charge up to 22 kW (32A, 3-phase).	**Sockets and plugs:** • **CHAdeMO socket.** Japanese standard for fast charging (DC). Can charge to maximum 65 kW per hour (see also appendix D). • **CCS socket.** German/American standard for fast charging (AC/DC). Can theoretically charge up to 170 kW per hour. Only one connection in the vehicle is used for both slow and fast charging. Suitable for smart grid applications. • **Tesla socket.** Standard of Tesla only, although the protocols have been released as open standards. Can charge up to 120 kW per hour. Suitable for smart grid applications. • **GB/T socket.** The Chinese standard is GB/T 20234 with a CAN bus communication protocol similar (but different) to CHAdeMO.

Source: http://www.chademo.com/wp/chademocharger/.

The Bhutan Electric Vehicle Initiative • http://dx.doi.org/10.1596/978-1-4648-0741-1

Table 6.4 Charging Modes

Charging mode	Communication	Locking	1-phase		3-phase	
Mode 1	No	In vehicle	Max. 16A	3.7 kW	Max. 16A	11.0 kW
Mode 2	Via charging cable and control box	In vehicle	Max. 32A	7.4 kW	Max. 32A	22.0 kW
Mode 3	Via charging station	In vehicle and charging station	Max. 63A	14.5 kW	Max. 63A	43.5 kW

Source: World Bank analysis.

Photo 6.1 Examples of Charging Stations

a. Mode 2 charging cable with built-in control box

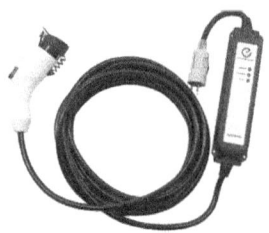

b. Charging station with multiple connections

c. ABB Fast charging station with multiple standards (CHAdeMO, CCS, AC)

Source: (a) © Nissan; (b) © EVConsult; (c) © ABB Ltd. All images used with the permission of respective copyright holders. Further permission required for reuse.

(OEMs) have so far obstructed a larger rollout of this technique. However, companies such as Better Place and Tesla have proved that the technique is possible and applicable to EVs. In Denmark, 17 Better Place battery swap stations (BSS) were realized outside of city centers and along major corridors (FDT 2013). Some of the advantages of these BSS are the very short charging times and the possibility to charge batteries during off-peak hours while using

> **Box 6.1 First Assessment: Charging Standards**
>
> Because of safety issues, it is recommended that EVs not be charged using standard household plugs. To determine which charging plug should be the standard (on the charging point side) and how technical and safety regulations should be adjusted in Bhutan, a testing and monitoring period of one year is recommended. The final decision for the standard plug is logically made in cooperation with China, India, and Japan because of the small size of the market in Bhutan. For fast charging, the charging standard is determined by the supply of vehicles, such as in the case of the CHAdeMO standard, which is now installed in Bhutan for the introduction of the Nissan Leaf. In the long term, multistandard fast charging is likely as a result of the different standards in China, Europe, Japan, and the United States.

them during peak hours. Negative aspects, however, include that (a) EVs with switchable batteries represent only a very limited share of the market and models with switchable battery are not on the OEMs' research and development (R&D) roadmap; (b) the investments cost are very high because the price of 1 BSS is similar to that of 750 normal charging stations; and (c) standardization of battery technology and switching techniques among OEMs has proved to be hard. Better Place filed for bankruptcy in 2013.

Induction. With induction charging, an EV can charge wirelessly by using an electromagnetic field to transfer energy between the charger and the EV. Similar to battery swapping, this type of charging has a high investment cost for infrastructure, requires standardization of battery packs, and involves issues concerning cooperation between OEMs. The technology also has a lower efficiency and increased resistive heating in comparison to conductive charging. Pilot projects, however, are showing that inductive charging is suitable for public transport. As the charging infrastructure can be installed at fixed locations such as bus stops, buses can charge each time for less than a minute (30–60 seconds) before moving on, with infrastructure placed at maximum intervals of a few kilometers.

EV Charging Options in Bhutan

The general charging requirements for Bhutan in terms of location—covering home, semipublic and workplace, public, and on-route charging—are the same as elsewhere. The context in Bhutan, however, differs from that in more developed EV markets, for example in terms of home ownership or possible EV uptake. This chapter describes the different possibilities for home, workplace, and fast charging in Bhutan. A special section addresses charging options for Thimphu because the largest and first uptake of EVs is expected in this city.

Home Charging at Private Homes, Apartments, and Public Locations

As discussed in "Types of EV Charging and Available Standards," above, home charging is essential for EVs because of the limited range of FEVs and the time required for charging (see also table 6.2). Home charging is relatively slow, and vehicles are usually charged overnight when the car is typically parked for more than 12 hours anyway.

Home charging (normally using 10–16 Ampere at 230V) requires a dedicated charging station, and the grid connection needs to be able to supply the required power (that is, 3.3–6.6 kW). With 3.3 kW of charging power, a station can charge a complete empty battery of a Nissan Leaf in about six hours. Some vehicles can also charge with 6.6 kW. In this case, additional connection and metering could be required because the maximum available power in a household is limited. Fast charging at home is not considered because of the high investment costs and low occupancy; it is also not necessary because of the long parking time.

Because EV owners in Bhutan cannot always park and charge their vehicles on private ground near their houses at night, scenarios for home charging (discussed below) cover not only use of a private driveway, but also arrangements at apartment complexes and in public areas. Additional information about parking and charging options for EV owners in Bhutan can also be collected in the planned transportation survey and be used to validate initial assumptions and plans for charging infrastructure. Photos 6.2 and 6.3 illustrate examples of charging stations for two home charging situations.

Home Charging Using a Private Driveway

When an EV owner can park and charge the vehicle on a private driveway, only an indoor or outdoor charging station has to be installed. If outdoor, a safe wall box is connected to the existing grid connection of the house, with a separate circuit breaker to protect the existing electrical installation. The station should be installed by a certified electrician and approved by a relevant entity, such as the Bhutan Power Corporation (BPC) and Thimphu Thromde. As mentioned, EV charging using an existing household socket is not recommended: such sockets are not designed for heavy use and have no separate grounding and circuit breakers in place.

Home Charging at Apartments Using Private Charging with Reserved Private Parking

At an apartment using private charging, EV owners can park their vehicles at home near the apartment building at clearly marked parking places reserved for EV owners. This reservation of parking spaces needs to be (financially) arranged with the landowner or other lessees and clearly indicated to prevent misuse of the parking spot by other vehicles.

The charging installation can be a wall box installed on the wall of the apartment building, which is connected to the private BPC grid connection of the EV owner. The EV owner finances and maintains the charger and pays for the additional electricity cost for charging through the regular electricity bill. The amount of electricity used for charging can be measured with a simple additional meter that is not a regulated meter from BPC. To ensure a safe operation, installation must be done by a certified electrician.

Charging Infrastructure and Network Planning 65

Photo 6.2 Private Charging Station with Socket for EV Plug

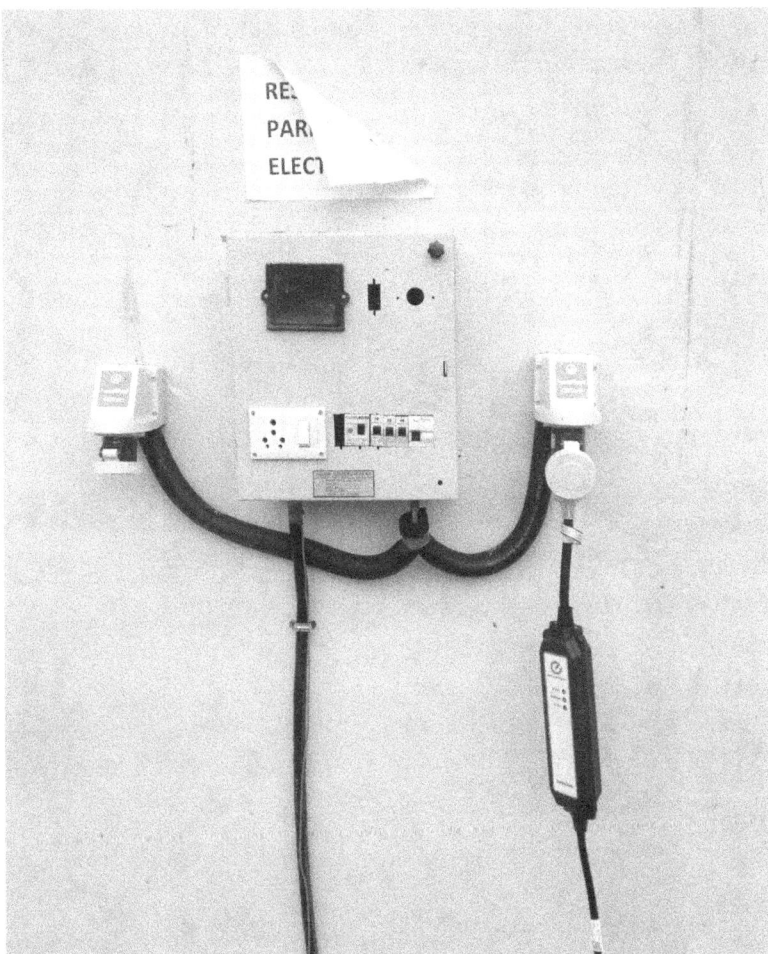

Source: © EVConsult. Used with the permission of EVConsult. Further permission required for reuse.

Because apartments in Bhutan will normally be equipped with a standard 60A single-phase meter (supplied by BPC to every household), the maximum load is limited to about 13 kW. In a normal situation, this should be enough to accommodate the charging of a standard Nissan Leaf at about 3.3 kW because an average household uses between 800 kWh per year in rural and 3,000 kWh per year in urban areas. However, when the charging is 7 kW (a charging option for the Nissan Leaf) and charging is done at peak moments (between 6 and 8 p.m.), the acceptable load on the circuit breaker for the apartment could be exceeded. BPC has indicated that additional measures will be required to move charging to off-peak hours or install a three-phase connection in the apartment.

The advantage of a setup with private parking and charging at an apartment building is the ease of use for the EV owner. A disadvantage, however, is that an agreement is needed from the apartment owner or lessor when the EV owner is

Photo 6.3 Outdoor Charging Socket at Clearly Indicated EV Parking Space

Source: © EVConsult. Used with the permission of EVConsult. Further permission required for reuse.

renting the apartment. In addition, when the EV owner leaves the apartment building, the EV charger is still connected to one specific home. Photos 6.4 and 6.5 illustrate possible reserved parking locations for home charging at an apartment. Box 6.2. provides a first assessment on charging at reserved private parking places.

In the future, provisions for EV charging can be included in building regulations.

Photo 6.4 Possible Private Parking Location at an Apartment

Source: © EVConsult. Used with the permission of EVConsult. Further permission required for reuse.

Home Charging at Apartments Using a Separate BPC Grid Connection and Reserved Private Parking

In the case of home charging at an apartment building without a private grid connection, EV owners also park their vehicles on a reserved private parking space at the apartment building, but the charging station is a (lockable) wall box or pole connected with a new and separate grid connection to the BPC grid. The separate grid connection expands the capacity and prevents a possible short circuit. The wall box or pole can be installed on or near the apartment building, which again should be done by a certified electrician with the setup approved by Thimphu Thromde and BPC and the apartment building owner. When more connections are available on one charger, an identification system is needed. The parking spaces near the charger should be reserved for EVs, with each EV having its own parking spot, or with several spots reserved for EVs in general.

If EV uptake is slow, a single charger will be sufficient for one apartment building. In a scenario with high EV growth, however, a double charger might be installed to accommodate future EV drivers. To determine individual electricity consumption, users could be provided with radio-frequency identification (RFID) cards, allowing the charging station's management system to provide separate billing. When multiple vehicles are connected, overloading of the grid connection can be prevented with an intelligent charging system that adjusts the charging load based on available power capacity. Intelligent charging, however, can currently only be done with mode 3 charging stations.

A separate BPC connection costs about US$180 (Nu 10,800), which, along with the supply and installation of the charging station, will need to be paid for.

Photo 6.5 Example of Reserved Parking Place for EV Charging

Source: © EVConsult. Used with the permission of EVConsult. Further permission required for reuse.

Box 6.2 First Assessment: Charging at Reserved Private Parking

For the low and high EV uptake scenarios, charging at reserved private parking places at apartment buildings is a viable solution that can be arranged by the EV drivers on a small scale. Building owners and lessees will need to be provided with incentives (or obligated) to allow chargers to be installed.

Charging Infrastructure and Network Planning

When more chargers are connected, the electricity and investment costs need to be split among the users. This can be a practical informal agreement with the lessees, comparable to the way water bills are split. A more scalable solution is to define a dedicated EV infrastructure operator to take care of the planning, installation, maintenance, and operations of the charging station (see also "Market Models for Ownership and Operation of Charging Infrastructure," below). These operations would also include invoicing services for charging and electricity cost. Box 6.3 provides a first assessment on reserved private parking and separate BPC grid connection.

Public Home Charging Using Reserved Public Parking and Extended Private Charging

In this version of home charging, the setup is similar to that at the apartment building but with the charging station placed in a public space instead of on private property (figure 6.2). The public charger is in this case used as a "public

---

Box 6.3 First Assessment: Reserved Private Parking and Separate BPC Grid Connection

For the super high EV uptake scenario with a large number of chargers, a separate grid connection with multiple sockets presents a scalable solution. With an average annual consumption of 1,900 kWh (kilowatt-hours), it is theoretically possible to charge up to six EVs. When all six EVs are connected at the same time, a three-phase connection is needed to allow for the load, which is then about 20 kW (6 × 3.3 kW).

Figure 6.2 Illustration of Extended Private Charging in Public Space

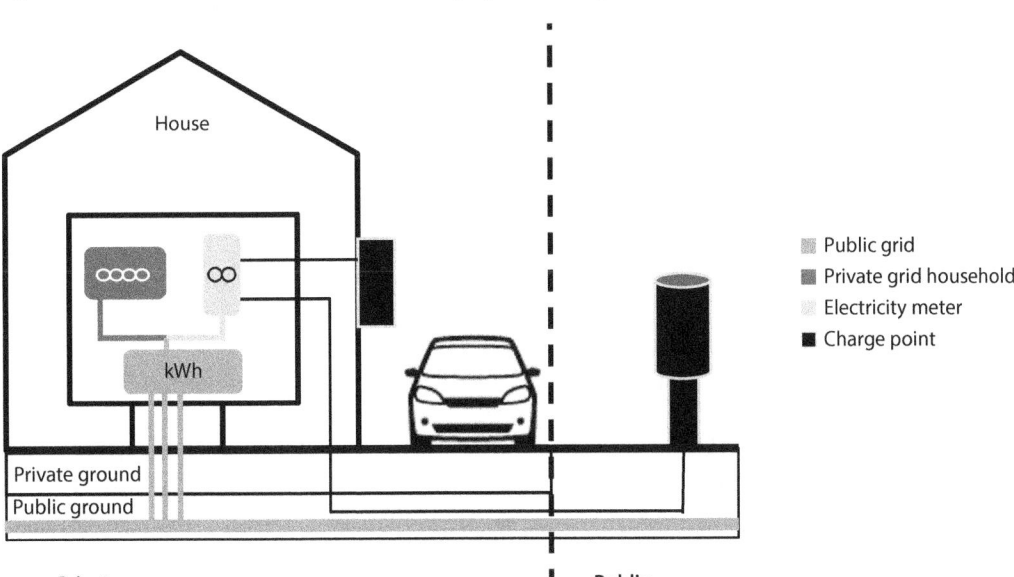

The Bhutan Electric Vehicle Initiative • http://dx.doi.org/10.1596/978-1-4648-0741-1

home charger." The charging station in a public space needs to be more robust and vandalism proof compared to those at homes and apartment buildings. In addition, the station needs to have Internet connectivity and an identification and payment system so EV drivers can find the charging point (for example on the Internet), identify themselves, and pay for the electricity they use. The public charger can be equipped with two sockets. In an area with many EV drivers, a setup with more connections can be installed.

For a charging station in a public area, an agreement with the city of Thimphu Thromde is needed about the ownership of the chargers, liability, insurance, maintenance, and reservation of parking spaces for EV charging. The city can also take care of the parking signs and parking enforcement to prevent nonelectric vehicles from parking at the public parking spaces intended for EV charging. Responsibilities for maintenance and operations also need to be clarified, and a dedicated infrastructure operator needs to be in place (see "Market Models for Ownership and Operation of Charging Infrastructure," below). Photo 6.6 shows examples of home charging with extended private charging. Box 6.4. provides a first assessment reserved public parking with extended private charging

Home Charging Using Public On-Street Parking and Slow Charging

EV owners without a private parking space or visitors from other areas could use public on-street slow charging stations to charge the EV overnight (see photo 6.7). Among all options for home charging, on-street parking with charging is the most complex because of financing and the large number of parties involved.

As shown in figure 6.3, involved parties involved typically include: (a) the supplier (and often producer) of the charging station hardware; (b) a back office providing the asset management tool for the chargers and performing remote monitoring, (c) a contractor to install (and sometimes maintain) the charging station, (d) the DSO responsible for the station's safe connection to the grid, along with managing the low, medium, and high voltage distribution networks, setting regulations, and approving new grid connections for charging stations; (e) the energy supplier, selling the electricity (either produced by the supplier or purchased on the energy market) to the charging station (Eurelectric 2013); (f) a maintenance and service provider to operate the charging station (including access control, management, repairs, helpdesk services, and billing) and provide (extra) services to the end user (Eurelectric 2013); and (g) the municipality or other party that is the legal owner of the ground and involved in (and often responsible for) the final decision for the location of a charging station and parking enforcement.

In the case of Bhutan, BPC would act as the DSO and be responsible for both the grid connection and energy supply, while the city could be responsible for parking (or parking concessionaires). Making BPC responsible can be organized at the national level by extending the regulated asset base to include the charging stations (for example with BPC as infrastructure operator). Otherwise, arrangements can also be made on the level of the city (or state, in the case of the United States), as is done in places such as Amsterdam, Berlin, and California (see also

Charging Infrastructure and Network Planning 71

Photo 6.6 Extended Private Charging Station with Two Sockets

Source: © EVConsult. Used with the permission of EVConsult. Further permission required for reuse.

Box 6.4 First Assessment: Reserved Public Parking with Extended Private Charging

For drivers without a private parking possibility, reserved public parking with extended private charging can be a solution as long as clear agreements about legal and insurance issues have been made with the municipality or ground owner.

"Market Models for Ownership and Operation of Charging Infrastructure," below). General requirements for public charging stations are summarized in box 6.5.

If an intelligent charger is used, identification and payment can be handled by an RFID card, used already in many countries worldwide. The EV driver registers as a user with the charging infrastructure operator and is supplied with a unique

Photo 6.7 Network of Charging Poles with 4 Sockets Connected to a Central Charging Hub

Source: © EVConsult. Used with the permission of EVConsult. Further permission required for reuse.

Figure 6.3 Parties Involved in Public Charging

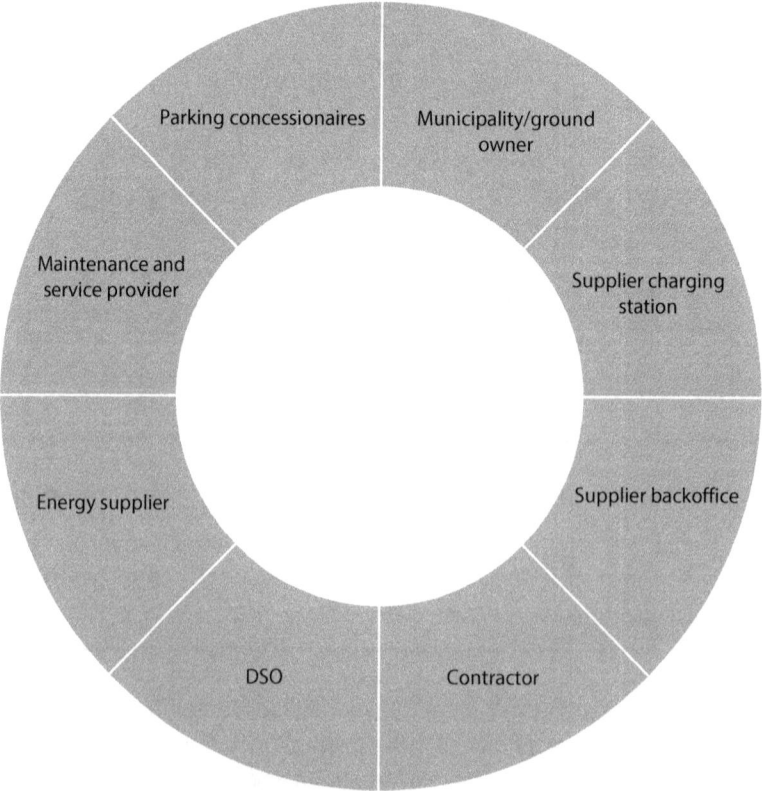

Source: World Bank analysis.
Note: DSO = Distribution System Operator.

Charging Infrastructure and Network Planning

Box 6.5 General Requirements for Public Charging Stations

Public charging stations in their setup need to be more robust than private charging stations. In general, in the setup of public charging stations the following eight aspects should be addressed:

- **Safety.** Safety measures are important because charging stations involve electrical work in possibly unattended public spaces, often in scarcely lit areas. The infrastructure operator is responsible for equipment safety.
- **Robustness.** Stations need to be designed for heavy use in public space and be vandalism proof.
- **Charging standards.** Charging standards and in particular the socket on the charger should accommodate all types of vehicles. The first chargers in Bhutan are now installed with the blue industrial socket (rated at 230V and 16A). (The charging station does not need to supply a charging cable because this is provided by the EV driver.)
- **Charging power.** For slow charging, a 230V and 16A connection is sufficient to charge normal EVs overnight.
- **Number of sockets.** A standard public charging station is equipped with one or two sockets. A master-slave combination is also possible, involving one intelligent charging station with small low-cost slave stations nearby.
- **Identification and payment.** Charging stations can be "simple" (just a safe socket on the street without a metering device or identification methods) or "intelligent" (a charger with methods for identification and payment).
- **Use of open protocols.** As more types and brands of chargers will be installed in the future, asset management will become increasingly complex. By using open protocols, such as Open Charge Point Protocol (OCPP), for intelligent chargers, stations can reduce their dependency on just one supplier and prevent a vendor lock-in.
- **Maintenance and service level.** The uptime of the charging network needs to be high to deliver reliable service to the EV drivers. If a charger is not working, this greatly inconveniences EV drivers.

RFID card. The user identifies at the charger with the card and starts the charging process. The charging station and ICT (information and communication technology) back office administer the use for every charging transaction and send a bill at the end of the month. This process is organized by the infrastructure operator. When an RFID card is used, the chargers need to be connected with a global system for mobile communication (GSM)/general packet radio service (GPRS) modem to an ICT back-office system, while each connection needs to have a dedicated meter to measure the energy consumption.

Providing public charging enables people without a private parking property to use EVs also, and the provision of public charging enhances the visibility of the EV program. However, as public charging is an expensive technical solution involving a complicated organizational model with multiple stakeholders, the business case for public charging is very difficult and public money is often required to ensure a healthy business case. Box 6.6 provides a first assessment for Bhutan and box 6.7 offers experiences in the Netherlands with public on-street parking.

Workplace Charging on Private Property or Public Locations

International experience illustrates that after home charging, workplace charging is the second most important method of EV charging. Workplace charging could take place at either private property near businesses or in public charging stations at strategic locations.

Box 6.6 First Assessment: Public On-Street Parking and Slow Charging

A first focus would best be on home and workplace charging because these are the most economic locations to realize charging infrastructure. Public charging might be provided only in a "last possibility" scenario, considering its high costs and involvement of a great number of stakeholders. Moreover, new parking concessions in the center of Thimphu have also made it more difficult to realize public charging stations because most on-street parking has moved to large parking garages.

Box 6.7 Case Study: Public Charging in the Netherlands

The municipality of Amsterdam, which has about 1,500 registered EVs and about 900 public charging stations, has invested US$6,5 million (Nu 390 million) in the hardware of public infrastructure (not including annual costs for service and maintenance). Most of the financing comes from the National Air Quality Cooperation Programme. This program was set up for an initial period of four years (2011–2015) to ensure the Netherlands would meet the deadlines for compliance with European Union (EU) limit values for particulate matter (PM10) and nitrogen dioxide (NO_2). In other regions in the Netherlands where money from the program was not available, investments were made by utilities and the private sector. Because of transformations in the market model and limitations in the amount of available public money, it is likely that in the near future the private sector will invest more in the public charging infrastructure of Amsterdam. It is significant that over the last four years the price of this hardware in the Netherlands has dropped in half.

Figure B6.7.1 depicts the development of public costs in charging infrastructure in the Netherlands. Public expenditure was calculated per year per double socket charge point at a consumer price of US$0.38 (Nu 22.8) per kWh including value added tax (VAT). These costs resulted from six public tender procedures in the Netherlands dating from 2008 to 2014.

box continues next page

Charging Infrastructure and Network Planning

Box 6.7 Case Study: Public Charging in the Netherlands *(continued)*

Figure B6.7.1 Development of Public Investments in the Netherlands

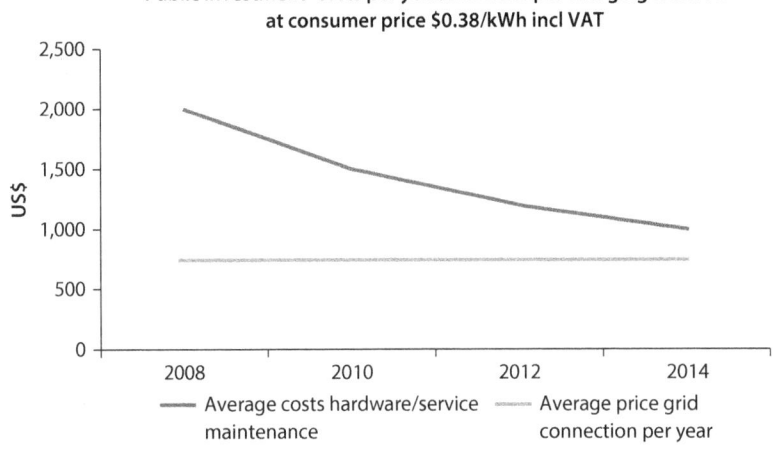

Source: World Bank analysis.
Note: kWh = kilowatt-hour; VAT = value added tax.

Grid connection costs are fixed because they are regulated. As total costs decreased, the share of the grid connection cost has therefore been growing steadily; as the business case slightly improved, less extra public investment was needed. Expenditure on the grid connection forms the "ceiling" in the business case. With the current regulatory framework and unchanged grid connection costs, the business case for private investors will not be positive without public investment.

Workplace Charging Using Private Parking

Workplace charging with private parking could involve EV charging offered by entities such as private companies, government institutions, government-related companies, hotels, and schools. Companies or institutions may supply the charging service to support the EV initiative but also for marketing purposes. This type of charging requires reserved parking spaces that are dedicated for EVs. Typically these parking spaces are at a visible location near the entrance of a building, which further serves as an incentive for employees to buy an EV.

The chargers for workplace charging will mostly be slow chargers because employees will normally stay at work for a longer period. The charger can be connected from the existing grid connection in the office building. Companies or institutions providing workplace charging for their staff and visitors could start with just a few (about one to four) chargers and expand as needed. If EVs are also purchased for company use, one charger needs to be installed for every vehicle.

Workplace Charging Using Public Parking

When employees cannot park and charge on private company property, they have to park on public streets and use a public charger on a reserved EV parking space to charge their vehicle during the work day. Public chargers for workplace charging can be installed at high-traffic locations where the chargers are expected to have a high occupancy, which will positively impact the business case. The requirements for public charging stations for workplace charging are the same as those for home charging using public on-street parking. See box 6.8 for a first assessment on public slow charging.

Public On-Route Fast Charging and Fast Charging for Taxis

While slow charging is sufficient and more affordable for home and workplace charging, fast charging is needed for range extension on longer drives and for taxi services, as taxis sometimes drive more than 100 km per day. With fast charging, a vehicle can be charged in 15–30 minutes to 80 percent state of charge (SOC). Several standards for fast charging have been adopted by vehicle manufacturers, and suppliers offer fast chargers with multiple standards (CHAdeMO, CCS, and AC) and chargers with multiple sockets of one standard. Over the next few years, fast charging operations are expected to change with new or adjusted standards and the availability of higher charging power.[2]

The advantages of fast charging—range extension and quick charging for taxis—come with a price: investment costs are high, a large grid connection (minimum 80A) is required, and the charging station has to serve multiple charging standards. The use of fast chargers is also hard to predict in advance. Moreover, of all the EVs, only FEVs will use the fast chargers, as PHEVs can use their own range extender running on gasoline. These factors combined make a business case for fast charging very difficult. In the Netherlands, the company Fastned has installed about 200 fast chargers throughout the country, charging about US$1 (Nu 60) per kWh.

Compared to slow charging, fast charging also must meet more advanced technical standards, which should be used when procuring equipment and operators. Box 6.9 presents a quick overview of requirements.

Planning for Fast Charging Locations for Range Extension in Bhutan

When using fast chargers, their geographical location should be carefully planned. Map 6.1 illustrates a first draft plan developed by BPC showing the location of 46 fast chargers along Bhutan's major route network. A phased rollout (illustrated

---

Box 6.8 First Assessment: Public Slow Charging

Public slow charging stations not related to a specific user should be installed only in high-traffic areas where a large number of EVs are expected.

Charging Infrastructure and Network Planning

Box 6.9 Technical Requirements for Fast Charging

Fast charging involves a high-level technical and electrical process between the charger and the vehicle. To deliver safe and reliable charging services, the equipment and operator need to be certified. When procuring fast charging equipment or contracting operators, some key elements that need to be addressed include:

- Applicable standards (for instance IEC 61851 for charging and IEC 61000-6-2/4 for Electromagnetic Compliance (EMC)) and potential adaptability at a later stage
- Number of charging connections
- Communication protocol with the EV
- Power per outlet and parallel power (20–50 kW)
- Protection (that is, IP54)
- Required input voltage (+range), current, and frequency
- Authentication (for example, RIFD, MIFARE, keypad, or button)
- Web-connection and asset management software with open standards protocol (OCPP, or Open Charge Points Protocol)
- Remote monitoring and update of functionalities
- Applicable installation altitude (up to 3,000 m), temperature, and humidity
- Efficiency and noise

Map 6.1 Draft BPC Plan for 46 Fast Chargers and Possible Phased Rollout

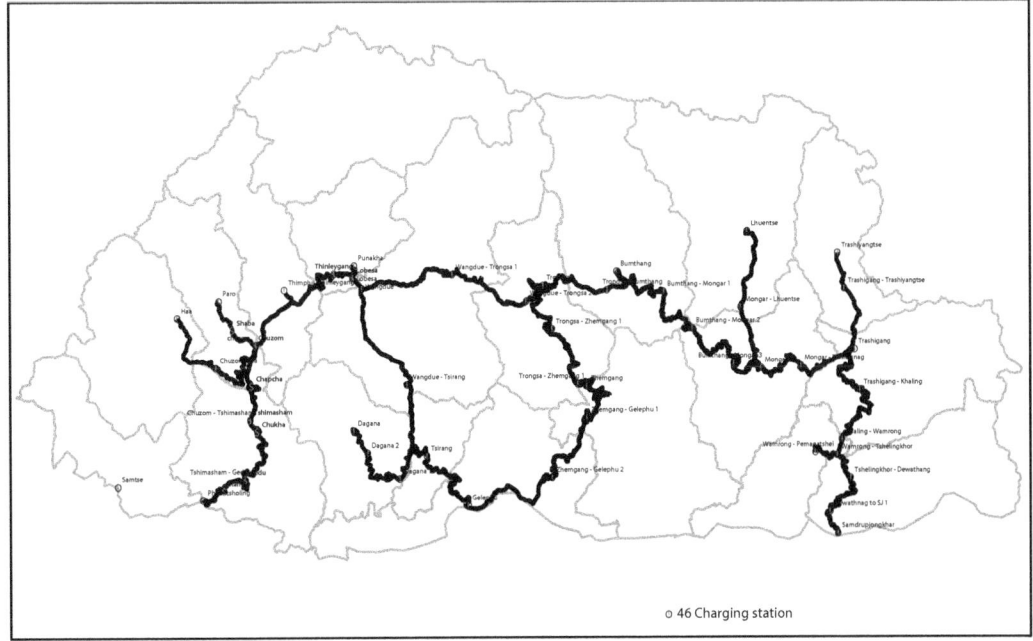

Source: BPC/World Bank analysis.
Note: BPC = Bhutan Power Corporation.

by the oval shapes in the map) can be used to slowly increase the number of fast chargers as EV uptake and demand grow. Rollout of fast chargers could start in and around Thimphu, where the major vehicle fleets and offices are concentrated, and at Paro airport for the first phase. The next phase, phase 2, can accommodate fast charging in Phuentsholing and Punakha region. A nationwide coverage of fast chargers is the final and ultimate stage 3 when the EV market is mature. From a customer perspective, a 30-minute charging event for roughly every 70–130 km of driving has a considerable impact on the driving experience. Box 6.10 provides a first assessment with some general remarks on fast charging in Bhutan.

When planning for fast charger locations, the following aspects should be considered:

- **Number of chargers per location.** Relevant international data are limited, but the rate of fast chargers per EV generally varies between 0.005 and 0.03 (or 1 charger for each 30–200 EVs) (Trigg and Telleen 2013). For calculation purposes, this report uses a rate of 1 fast charger per 100 normal EVs, which takes into consideration that public home charging is hard to realize. This means that in an area like Thimphu with 1,000 EVs, about 10 fast chargers are required, not including the fast chargers required for specific target groups such as taxis.

- **Location of range extension fast chargers.** Fast chargers aimed at range extension should be located along major traffic routes and as close to the road as possible, accessible from both directions. Locations close to a junction are very suitable. The location should also be comfortable for a 15–30 minute wait for the EV drivers and passengers.

- **Distance to the electricity grid.** An important factor for fast charger locations is the relative distance to the electricity grid. As one 50 kW DC charger needs a 400V three-phase 90A/100A grid connection, a location with five chargers will need a 5*50 kW = 250 kW or 300 kVA (kilovolt Ampere) input to operate the five chargers at maximum power simultaneously. This requirement can be lowered if it is accepted that the power output is lower (and charging time

---

Box 6.10 First Assessment: Fast Charging

EVs in general make short trips, and fast charging for range extension is used only incidentally. The rollout of the fast charger network can grow in parallel to the uptake of EVs in the country, or just slightly ahead to convince consumers, with the number of fast chargers growing gradually with the number of EVs. Installing a nationwide fast charging network is no guarantee for fast EV uptake because other aspects, such as incentives, EV awareness, and availability of suitable models, also need to be in place.

higher) when all chargers are occupied simultaneously. A detailed assessment of highway fast charging locations in relation to the electricity grid is needed. The number of chargers per location and subsequent power input requirements both depend on the EV uptake scenario in Bhutan.

- Air density. Because of the altitude of some parts of the country, the fast chargers' protocols will need to be adjusted to accommodate for lower air density. Most common fast chargers are guaranteed up to a height of about 1,000–1,500m.

Fast Charging to Increase the Operational Capacity of Taxis

EV taxis will need fast charging during the day to be able to provide normal services. A taxi driving 50,000 km a year over 250 working days drives an average of 200 km a day. As the expected range for an EV is about 120–150 km, at least one fast charging event is needed during the day to be fully operational, assuming the taxi can start each day with a full battery charged with a (public) home charger. With fast charging, the EV taxi can be charged to 80 percent SOC within 30 minutes, which could be done during a break. Slow charging (two to four hours) is not acceptable for taxis during the day because of the negative impact on the taxi business case.

In planning for fast charging for taxis, the following considerations apply:

- **Number of fast chargers per taxi vehicle.** This number is related to the occupancy of the charger. A good starting point for providing fast chargers is having 20 taxis per single charger. If this single charger is operated for 13 hours a day (assuming a work day from 7 a.m. to 8 p.m.), it can theoretically serve 26 vehicles with a charging time of 30 minutes. However, because this would give the charging station an occupancy of 100 percent during working hours, this level of occupancy can be achieved only with a row of waiting taxis. If occupancy is 75 percent during working hours, the charger can serve 0.75*26 = 20 vehicles during the day. This calculation can change if, for example, charging time is less than 30 minutes, more taxis are waiting, or operating hours are extended.

- **Charger occupancy and use during the day.** To prevent high investment costs and low occupancy, the use of taxi fast chargers should be spread out over the day. A mobile phone app as a planning and reservation tool could be used for this purpose.

- **Location of the charging hub.** The fast charger should be as close as possible to the user group. For taxis in Thimphu, for example, this charging hub has to be on or near the taxi stand. A drive-by taxi stand or pick-up point along the road is not suitable for fast charging because vehicles have to keep moving toward the front position to pick up passengers and thus are parked for only a few minutes at a time. In addition, from a city planning and transport perspective, the charging hub and surrounding roads need to be able to accommodate the additional traffic. The charging location should also have enough area available to install the charging equipment and transformer station. Photos 6.8 and 6.9 illustrate examples of a suitable and nonsuitable location for this kind of charging.

Charging in Thimphu

The uptake of EVs is expected to be the highest in Thimphu, and this section summarizes the current situation and plans for public charging stations in

Photo 6.8 Examples of Taxi Stands Not Suitable for a Charging Location Because of Lack of Space

Source: © EVConsult. Used with the permission of EVConsult. Further permission required for reuse.

Photo 6.9 Example of Possible Charging Location for Taxis in Thimphu with Sufficient Space and Bhutan Power Corporation Connection

Source: © EVConsult. Used with the permission of EVConsult. Further permission required for reuse.

the city. Private home and workplace charging in Thimphu will be similar to the situations described for Bhutan in general.

The City of Thimphu is organized into about 12 neighborhoods. For fast charging, the already planned "neighborhood nodes," as described in the Thimphu Structure Plan 2004, may be a suitable location because they are centrally located and planned around a concentration of local facilities, such as shops, police stations, banks, and schools. Thimphu Thromde already owns plots of land in all neighborhoods to develop these nodes. While some nodes have already been developed, others may take a few more years. For public normal charging, however, the nodes are not as suitable because the walking distance from the neighborhood node to the driver's home will be quite large. Although little international data are available on what is considered the maximum distance from charger to home, it is expected that for everyday use a walking distance of about 300 meters is acceptable. Although a general statement is difficult to make with little information available, it is expected that in Thimphu about 70 percent can find a way to park on private ground.

Depending on the EV uptake, different numbers of fast chargers would need to be installed. For all scenarios, however, installation could start with locations in the very north, south, and city center of Thimphu because these would be the most convenient for all drivers. A detailed plan for the exact number and type of chargers at each location will need to be developed when a scenario has been chosen. Thimphu Thromde has done an initial assessment for 20 possible locations for fast chargers, some of which have been visited and assessed for charger installation (appendix E).

Other plans for charging in Thimphu include equipping the planned new parking garages in the city with a number of normal chargers and reserved parking places for EVs. To create an incentive, the possible public parking spaces for EVs could be free parking. The city, however, has not yet reserved budget for such a measure. In general, the city council of Thimphu Thromde has not yet actively planned or prepared for charging stations. The city has, however, indicated that one of its main concerns is that unmanned electrical equipment like charging stations on public streets will be vandalized or demolished. This is based on experience with a high rate of broken automated teller machines (ATMs) and demolished street lighting.

Charging Infrastructure Requirements by Uptake Scenario and National Rollout

The sufficient availability of robust charging stations is important for EV uptake, EV driving, and consumer confidence in the technology; but, because of the high investment costs, infrastructure development and rollout should be carefully planned to prevent low returns on investment and ensure high occupancy of the chargers. As the investments needed also depend on the uptake of EVs, an accurate and credible assessment of likely uptake is essential to plan for the required investments.

Using the three EV uptake scenarios defined in chapter 3, the tables in this section summarize the required number of chargers for each target group (government fleet, taxis, and private vehicles), based on assumptions for the distribution of EV uptake among these target groups. Real uptake is also influenced by other factors. The availability of EV chargers is a prerequisite for EV uptake, but no guarantee as uptake also depends on EV awareness and incentives, among others.[3]

The calculations assume that 70 percent of EV owners can find a way for home charging on private ground, even if that includes a private arrangement at a location near the home—leaving 30 percent of drivers in need of a public slow charger (see box 6.11 for a list of all assumptions). The required number of chargers for each scenario and target group is then based on the number of EVs for each scenario (see also "Three Scenarios for EV Uptake in Bhutan" in chapter 3)

Box 6.11 General Assumptions for the Calculation of Charging Infrastructure Requirements

To calculate the need for charging infrastructure, calculations in this report use the following assumptions:

- **One home charger per vehicle.** Every vehicle requires one single slow home charger, ensuring all EVs can start with a full battery in the morning.
- **70 percent home charging on private ground.** About 70 percent of EV owners can find a way for home charging on private ground. This can be directly at the driver's home/apartment or through a private arrangement in the vicinity of the home. EV pioneers tend to be creative about finding charging solutions.
- **30 percent home charging using public slow charging.** About 30 percent of EV drivers have to use a public slow charger because they can park only on public streets and have no access to private charging near their homes.
- **Only slow charging for home charging.** The use of fast chargers as a primary charging method (without home charging) is not considered as it affects the driving pattern too much.
- **Twenty taxis per fast charger socket.** One fast charger socket can serve 20 taxis per day (using a charging time of 30 minutes and a charger operation of 10 hours per day). This one charger can be one socket on a multiple socket charging station.
- **One hundred private or government vehicles per fast charger.** One fast charger (one socket) can serve 100 private or government fleet vehicles. If a driver charges once a week for 30 minutes, and one socket can serve 20 drivers per day, the total number of users is 100 for 5 working days. This rate of 0.01 fast chargers per EV is the global average (International Energy Agency 2013).
- **Two private or government vehicles per workplace charger.** Every private and government vehicle requires 0.5 workplace charger. One charger for two vehicles is sufficient because a vehicle is not always parked at the office and drivers can share an office charger. For calculation purposes, these chargers are assumed to be all private chargers.

Charging Infrastructure and Network Planning

and the assumptions about charging; this, however, is only a rough calculation to give an indication of the required investments for a charging network. The assumptions will need to be monitored and adapted to the outcome of pilot projects and ongoing international experience.

Low EV Uptake Scenario

In the low EV uptake scenario, by 2020 about 648 normal chargers (for home, work, and public charging) and about 10 fast chargers will be required (table 6.5). For this scenario, it is assumed that uptake will mainly take place in the Thimphu area. The number of fast chargers is calculated only in relation to the number of EVs, and no nationwide coverage by fast charging is taken into account.

The low EV uptake scenario does not require a large coordinated approach among the stakeholders. The public chargers might be installed by apartment owners, while operation of fast chargers can be done by a small-scale company. No special rules or adjusted regulations are needed to accommodate this scenario.

High EV Uptake Scenario

In this scenario, the number of fast chargers is calculated in relation to the number of EVs, with another 10 fast chargers added to cover the third phase of geographical coverage (covering Punakha, Chukha, Gedu, and Phuentsholing, see map 6.1). These additional 10 fast chargers should be placed along the highways to accommodate on-route charging and thus extend the range; from a capacity perspective this number of chargers is not yet required for this scenario. As shown in table 6.6, by 2020 a total of 1,988 normal chargers and 43 fast chargers are required.

In this scenario the public charging network (slow and fast) is an integral part of the EV strategy and a charging operator (see "Market Models for Ownership and Operation of Charging Infrastructure," below) needs to be in place from

Table 6.5 Low EV Uptake Scenario: Estimated Number of Chargers Required by 2020

		Number of chargers			
	Number of EVs	Home private	Home public	Work	Fast
2015					
Taxi	22	15	7	0	2
Government	3	2	1	2	0
Private	54	38	16	27	1
Total 2015	79	55	24	29	3
2020					
Taxi	132	93	40	0	7
Government	21	15	6	10	0
Private	323	226	97	161	3
Total 2020	476	334	143	171	10

Source: World Bank analysis.
Note: EV = electric vehicle.

The Bhutan Electric Vehicle Initiative • http://dx.doi.org/10.1596/978-1-4648-0741-1

Table 6.6 High EV Uptake Scenario: Estimated Number of Chargers Required by 2020

		Number of chargers			
	Number of EVs	Home private	Home public	Work	Fast
2015					
Taxi	74	51	22	0	4
Government	10	7	3	5	0
Private	161	113	48	81	2
Total 2015	245	172	74	86	16 (6 + 10 for range extension)
2020					
Taxi	441	309	132	0	22
Government	62	44	19	31	1
Private	969	678	291	484	10
Total 2020	1,472	1,031	442	515	43 (33 + 10 for range extension)

Source: World Bank analysis.
Note: EV = electric vehicle.

the start. Reliable operation of the charging network is essential to operate the taxi fleet and private vehicles.

Very High EV Uptake Scenario

In the scenario with very high EV uptake, each year about 1,000 EVs enter the market. These vehicles require all charging facilities to be fully operational from day one. As described below (figure 6.4), the rollout of a nationwide charging network will take at least 30 months. This scenario will involve a rollout in many cities at the same time, with a large focus on the Thimphu area because the majority of vehicles would be concentrated there. Table 6.7 lists the numbers of required chargers. Under this scenario, no additional fast chargers are needed for range extension.

The numbers of fast chargers in the table are the number of sockets (connection for one car). For example in 2020, 221 fast chargers are needed for taxi use across Bhutan. How these fast chargers are distributed across the country and within Thimphu needs to be further assessed. If about 70 percent of the chargers will be in Thimphu, this would cover 167 chargers. Within Thimphu, the main three taxi stands could then, for example, each have 20 double socket fast chargers, for a total of 120 available sockets across the three stands. The remaining 47 fast chargers for Thimphu could then be distributed among the 12 neighborhoods. The 30 percent of chargers used outside Thimphu should be distributed across the country.

Phased Rollout and Timeline for the Development of the Charging Network

The development and rollout of a nationwide charging network is a complicated process that requires a dedicated organization, sufficient financing, and technical expertise. To organize the rollout, the roles of the involved stakeholders and

Figure 6.4 Indicative Plan and Timetable for the Realization of the Charging Network

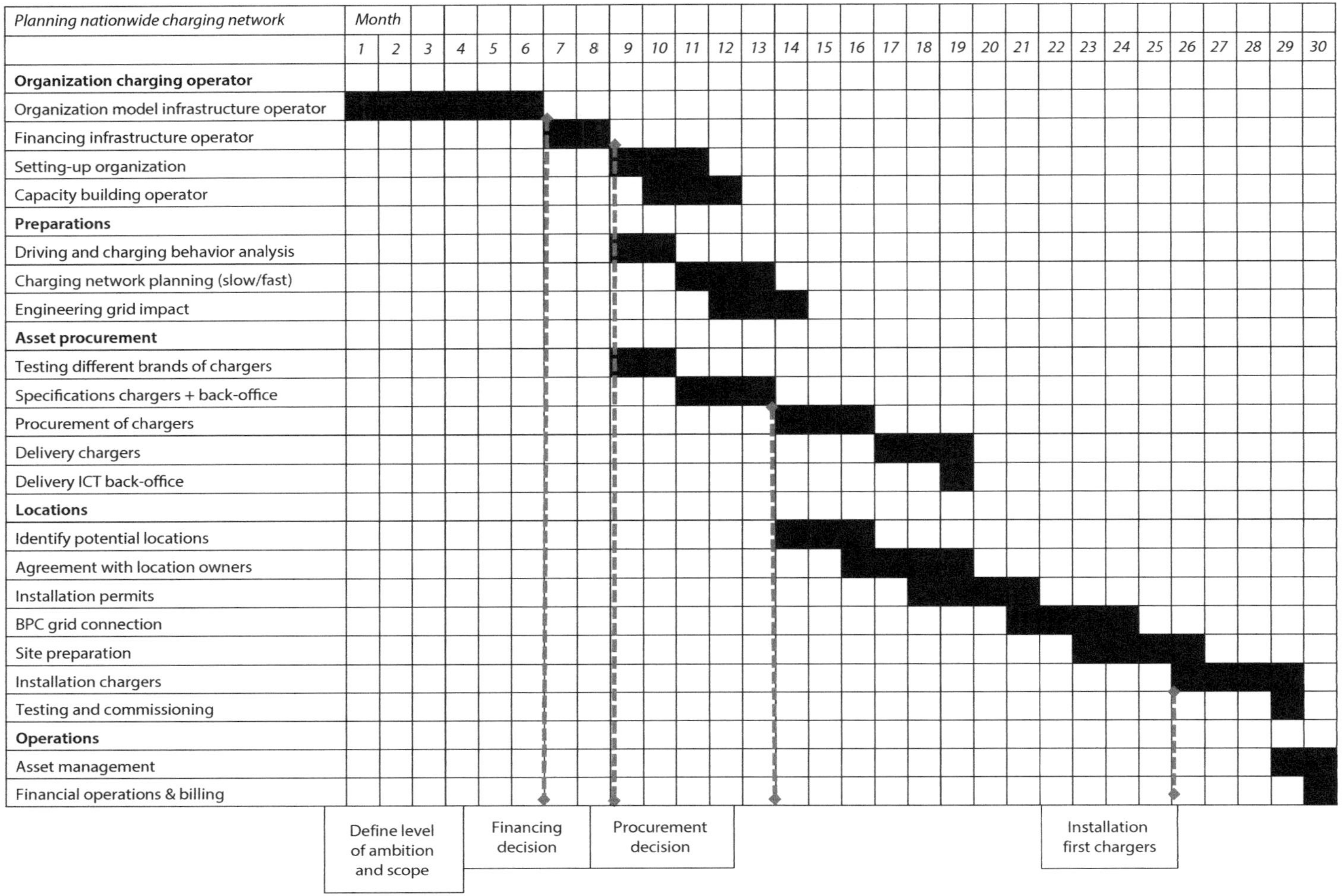

Planning nationwide charging network	Month																													
	1	2	3	4	5	6	7	8	9	10	11	12	13	14	15	16	17	18	19	20	21	22	23	24	25	26	27	28	29	30
Organization charging operator																														
Organization model infrastructure operator																														
Financing infrastructure operator																														
Setting-up organization																														
Capacity building operator																														
Preparations																														
Driving and charging behavior analysis																														
Charging network planning (slow/fast)																														
Engineering grid impact																														
Asset procurement																														
Testing different brands of chargers																														
Specifications chargers + back-office																														
Procurement of chargers																														
Delivery chargers																														
Delivery ICT back-office																														
Locations																														
Identify potential locations																														
Agreement with location owners																														
Installation permits																														
BPC grid connection																														
Site preparation																														
Installation chargers																														
Testing and commissioning																														
Operations																														
Asset management																														
Financial operations & billing																														

Milestones: Define level of ambition and scope | Financing decision | Procurement decision | Installation first chargers

85

Table 6.7 Super High EV Uptake Scenario: Estimated Number of Chargers Required by 2020

	Number of EVs	Number of chargers			
		Home private	Home public	Work	Fast
2015					
Taxi	736	515	221	0	37
Government	17	12	5	9	0
Private	269	188	81	135	3
Total 2015	**1,022**	**715**	**307**	**143**	**40**
2020					
Taxi	4,414	3,090	1,324	0	221
Government	104	73	31	52	1
Private	1,615	1,130	484	807	16
Total 2020	**6,132**	**4,293**	**1,840**	**859**	**238**

Source: World Bank analysis.
Note: EV = electric vehicle.

other existing organizations (such as automotive organizations, the energy company, and city council) need to be defined. Moreover, a dedicated charging network operator (see "Market Models for Ownership and Operation of Charging Infrastructure," below) should be identified or set up. Because this is a new role, expertise may have to be brought in.

Along with the organizational preparation, the technical requirements for slow and fast chargers need to be defined and a procurement process put in place. Locations for charging stations have to be identified, and permit procedures need to be followed. The locations also have to be prepared for installation. The BPC grid connection has to be installed where lead-time is dependent on distance to the existing electricity network. In addition, EV charging is a new field and is often not covered in existing regulations on safety, electricity installations, house wiring, and standards. The lessons from a pilot period can deliver input to change these regulations in the future.

Figure 6.4 indicates a possible timeline for the rollout of a complete charging network, covering the setup of the organization of the charging operator, preparations, asset procurement, the preparation of locations, and operations. In reality, the implementation of the charging network will be a gradual process with some chargers installed early and others much later, for example if a good location is not available or because of delays in the installation of the grid connection.

The introduction of EVs will be done in phases and the rollout of chargers has to follow these phases. Not all information is available right at the start. Every next phase can be adjusted with lessons from the previous one. The first phase will be a pioneering phase with delivery of first vehicles and chargers organized mainly by a small group of enthusiasts. The next phase can be an orchestrated learning and pilot phase. The phases of scaling up and mass adoption will follow later, when the aggregated learning can be used to efficiently roll out the charging networks.

Market Models for Ownership and Operation of Charging Infrastructure

Market models describe the interactions, roles, and responsibilities for the different players in the market for providing public charging infrastructure, covering the distribution of electricity, the operation and ownership of the charging station, retail of energy, and e-mobility services and products (Eurelectric 2013). A market model is different from a business model in that it does not provide information about the revenue streams of individual business players. This section introduces international experience with market models and presents the situation in Bhutan.

International Market Models

Two main market models can be distinguished for the ownership and operation of EV charging infrastructure. These are (a) an integrated infrastructure model (also known as the integrated DSO model) used in China and Ireland, and (b) an independent e-mobility model, such as those used in the United States.

Model 1: Integrated Infrastructure Model. In the integrated model, the charging infrastructure is integrated in the DSO's (regulated) activities and asset base. As shown in figure 6.5, the DSO in this model is responsible for the distribution of the energy and also operation and maintenance of the charging stations. Retail of energy and services can be provided by the DSO but also by independent (new) market parties. The costs for investments and operation are shared by EV users and drivers of regular vehicles via the energy and network use tariffs. The DSO secures the stability of the grid and safety of installations since it controls the charging process. In case multiple DSOs have deployed (different kinds of) public charging infrastructure, a "clearing house" is needed to clear business-to-business (B2B) settlements concerning financial, information, and billing issues. In Bhutan, the DSO is BPC.

Model 2: Independent e-Mobility Model. In this model, used in Germany, the Netherlands, and the United States, market parties deploy the charging stations independent from the DSO (figure 6.6). The DSO has the regulated task to connect the charging point with the grid, just as any regular household, but several market parties run the activities to provide the e-mobility user with power. This includes the installation and operation of the charging station, ICT and back office of payment systems (possible with roaming agreements and credit card

Figure 6.5 Integrated Infrastructure Model

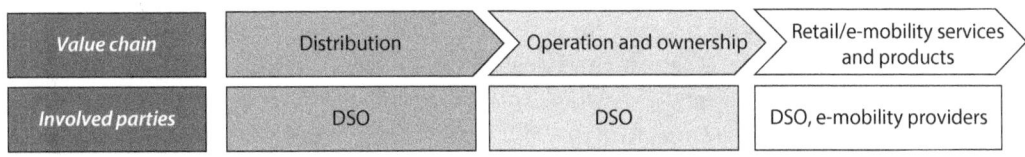

Source: World Bank analysis.
Note: DSO = Distribution System Operator.

Figure 6.6 Independent e-Mobility Model

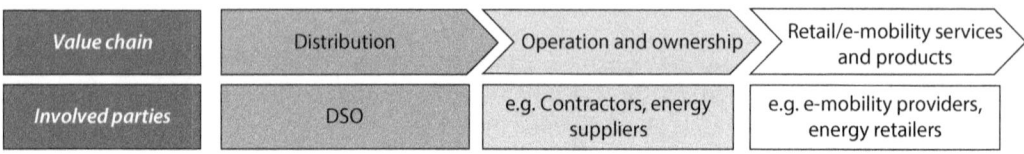

Source: World Bank analysis.
Note: DSO = Distribution System Operator.

Table 6.8 Advantages and Disadvantages of the Two Market Models

	Advantages	Disadvantages
Integrated infrastructure model	• Structural and guaranteed rollout of safe (public) charging stations • Safety and reliability guaranteed. • Affordable charging tariff. • Control over grid for DSO. • Integrated in "regular" DSO activities.	• Costs socialized via taxes, also for non-EV drivers. • Little/no competition and innovation. • Rising charging tariff in the (nearby) future because of difficult business case. • Monopoly of DSO.
Independent e-mobility model	• Encourages innovation and competition, development of new companies. • Multiple options for EV drivers. • Little/no government money required (but subsidies can be used to stimulate the market).	• High charging tariff and little government control over tariff. • Control required on charging market development. • Difficult business case. • Smart charging applications needed for maintaining safe and stable grid. • Multiple types of charging points in one street, area, or town.

Source: World Bank analysis.
Note: DSO = Distribution System Operator; EV = electric vehicle.

payments), and products and services to the e-mobility users. Although various versions of this model exist, the infrastructure operators are always responsible for generating their own revenues from selling electricity to run the operations. To stimulate the market, the government can provide subsidies to the infrastructure operators to encourage growth of the network and overcome the negative business case at the start-up period.

Table 6.8 presents an overview of the advantages and disadvantages of the different models. When organizing EV charging infrastructure, countries may adapt either approach, transition from one to the other, or develop a hybrid model using aspects of both. Table 6.9 presents an overview of the way the market models are used in various countries.

Role for the Charging Operator

In both models, the role of the charging and infrastructure operator is key to developing a charging network for both public slow and public fast charging. The operator takes care of all activities needed for a reliable network, including planning, preparation, supply, installation, operations, billing (often in combination with a service provider), maintenance, and repairs. The operator also arranges agreement with the location owner, such as for example the dzongkhag or a private location. Box 6.12 illustrates a few operations of the charging operator back office.

Table 6.9 International Market Models

Country	Finance	Organization
Ireland	DSO	• Integrated infrastructure model (DSO) • Publicly accessible charging infrastructure is part of the regulated business of operating and managing the low voltage and medium voltage electricity grid. • Roaming of electricity and service by DSO. • DSO has ICT back-end system for asset management. • Customer linked to e-mobility service provider. • E-mobility service provider has access under nondiscriminatory conditions.
Netherlands	Government, private sector	• Started with the integrated infrastructure model. After about 3 years, in a more mature market, a shift was made to the independent e-mobility model. • Public charging points are deployed independently from the "regulated" DSO/grid business. • Over 1,500 public charging points are realized by market parties. • New connection points for public charging stations are treated as any other new connection points to the distribution grid. • When customers want to use a charging station operated by a different e-mobility service provider, access can be granted via a roaming agreement. • Nissan donated 100 DC fast charging stations from DBT-CEV, but these have been removed because of poor quality and unreliability.
China	DSO (state grid, China Southern grid)	• Integrated infrastructure model (DSO). • Plans to build more than 2,350 EV charging stations and 220,000 charging poles by 2015. • Promotion of battery swapping. • China Southern free-of-charge installs two EV charging poles for each Shenzhen EV driver, one at the home or apartment of the driver and another near or at the driver's place of business.
Korea, Rep.	Government, private sector	• The Ministry of Environment (MOE) has completed building 80 high-speed charging stations for public use to expand distribution of EVs. • Chargers are operated by the Vehicle Environment Certification Centre of Korea Environment Corporation (KECO). • The public charging stations will be provided for free. A charge will be applied for using stations from 2014 after the pilot period. • As part of BMW's recent efforts to support electric vehicle adoption around the globe, BMW donated 30 charging stations to Jeju Island.
United States	Market parties, subsidies from public sector	• Independent e-mobility model. • Energy companies are still largely uninvolved in the emerging EV charging market, partly because they are limited by regulation. • Investment by the U.S. Department of Energy (about US$800 million or Nu 48,000 million) for several projects. An example project is Ecotality (US$99 million or Nu 5,940 million), which sells electric vehicle supply equipment (EVSE). • Eco Texas: NRG eVgo is a privately funded part of NRG Energy Inc. NRG eVgo's investment of approximately US$150 (Nu 9,000) million provides hundreds of public fast charging sites along with individual charging stations at homes, offices, multifamily communities, schools, and hospitals across networks in Texas, California, and the Washington, DC, Maryland, and Virginia region. • The U.S. federal tax code provides incentives for the purchase and installation of EVSE. • California and Massachusetts have both authorized state subsidies for EVSE, albeit at a pilot scale. Michigan's Local Distribution Company offers rebates for residential EVSE.

Source: World Bank analysis.
Note: DSO = Distribution Service Operator; EV = electric vehicle; ICT = information and communication technology.

Box 6.12 Operations of a Charging Operator Back Office

In general, costs of operations and maintenance can be greatly reduced if a back-office system is used to remotely access and monitor chargers. Such a back-office asset management system, as for example shown in figure B6.12.1, can provide information on the status of the charger and relevant events, such as charging time, disconnections, and errors. If needed, the chargers can also be reset remotely. To invoice users for the charging service, charging stations need to be equipped with a method to identify users, such as a key, a pin code, or RFID card. For RFID identification, a mobile connection is required, which might be difficult in remote areas. Manual invoicing is considered to be very unpractical.

There are various types of charging infrastructure that can be combined with separate back-office systems for asset management and customer management. The customer management part is for registration, user information, and billing. International companies like ABB, CGI, IBM, Siemens, Robert Bosch, Chargepoint, and Greenlots offer these back-office services.

Figure B6.12.1 Screenshot of Back Office Charging Operator

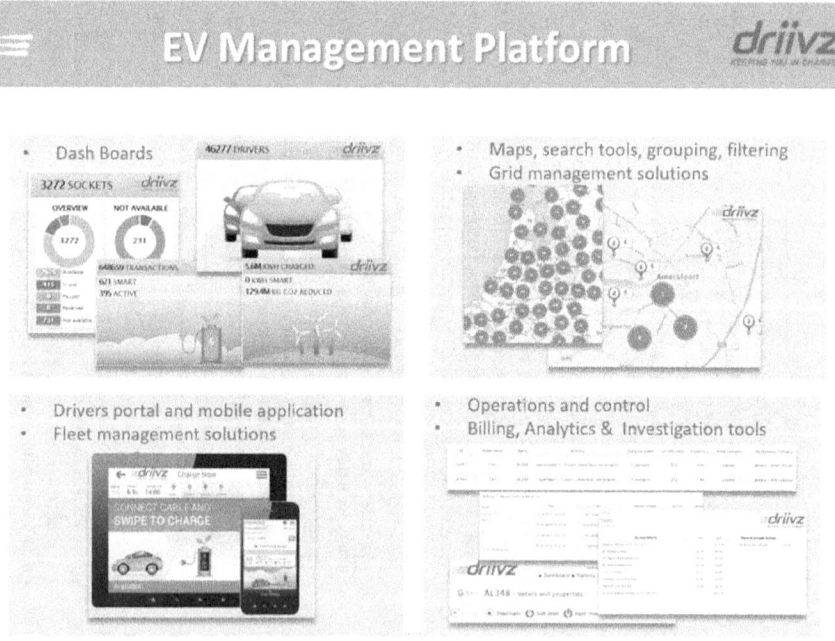

Source: © Driivz Ltd. Used with the permission of Driivz Ltd. Further permission required for reuse.

In different countries, the role of the charging operator is picked up by different kinds of organizations. Traditional contractors or large infrastructure producers (such as ABB or Bosch) see the installation and maintenance of charging points mainly as an expansion of their regular business activities. Specifically in the area of customer relations, however, new companies are developing. These companies are using a new business model that is focused on taking care of the whole process, from the request of a charging point and charge pass to the billing of invoices. These companies fill the gap left by traditional car dealers.

In Denmark, for example, CLEVER has successfully positioned itself as the new e-mobility service provider in the market. The company offers charging solutions to companies for (commercial) exploitation and is working closely with leasing companies and OEMs. A quick rise of these kinds of companies in several countries is due to the fact that traditional parties, such as oil companies and OEMs, do not see the EV sector fitting into their regular business activities. However, examples also exist of more traditional companies picking up these roles. In the United Kingdom, the already existing company British Gas (an energy and home service provider with more twelve million customers) installs, maintains, and is a service provider for charging stations.

Charging Infrastructure Market Model for Bhutan

Each country has its own approach to a market model for infrastructure development (see table 6.9); and, because EV is a new sector in Bhutan, the roles and responsibilities of the stakeholders have not yet been defined. The decision about the market model for charging infrastructure is a decision by the national government, based on available budget and on policies for energy market development, economic development, private sector development, and EV policy.

In Bhutan, BPC is the regulated grid operator. The corporation has experience in the electricity market and is involved with long-term grid planning as well as household-level electricity connections. Although it is well suited to be the charging operator and do the rollout for the home, public slow, and public fast charging infrastructure, BPC has indicated this role is outside the boundaries of its existing mandate and that it will support only a designated charging operator with technical support, installation of the grid extension, installation of transformers, and electricity supply. The selection of a charging infrastructure operator will be an important first step for the roll out of charging infrastructure in Bhutan (box 6.13).

In addition to the BPC, the RGoB itself will need to play a key role. The supply of chargers is currently envisaged to be the responsibility of the RGoB. A detailed set of requirements for this supply needs to be prepared, describing the technical standards, number and power of connectors, safety requirements, adjustments for installation at higher altitude, requirements for the payment and asset management system, adaptability to future standards, maintenance contract, and availability of spare parts. Possible operators could also be involved in the procurement process to make sure the chargers are suitable for commercial operations. The RGoB has received a number of fast charging stations from car manufacturers like Nissan, which can be supplied to a charging operator under certain conditions.

Box 6.13 First Assessment: Selecting a Charging Infrastructure Operator

As BPC has indicated the role of charging operator is outside its boundaries, a dedicated commercial operator will need to be selected to assure safe and reliable charging options for EV drivers. Preparations for the selection of a charging infrastructure operator should include the following:

- **Development of a detailed scope of work for the infrastructure operator.** A detailed description of the operator's scope of work is needed, covering for example the installation of slow and public fast charging stations, maintenance, operations, and marketing. (The supply of land, grid installation, and supply of fast chargers is initially envisaged to be a responsibility of the RGoB).
- **Agreement with BPC on financing and the scope of BPC support to the operator.** BPC support should include grid extension, installation of transformers, and technical support. BPC needs an agreement from the Bhutan Electricity Authority (BEA) to extend the grid, own the additional assets, and possibly recover these costs.
- **Definition of clear terms and conditions for using the land.** If land is used from, for instance, the National Land Commission or Thimphu Thromde, the terms and conditions for this use should be clear. The locations for the chargers can be preselected by the RGoB with consultation from relevant stakeholders. International practice suggests supplying the land at zero cost for a fixed period to support the starting companies.
- **Development of a widely supported EV policy and EV uptake scenario.** The RGoB should also indicate the duration of the incentive programs. Based on the policy and desired uptake scenario, the location and number of fast chargers can be planned. The information will also help commercial parties understand the future business case.
- **Understanding of the operator's business case for the initial and mature operation phases to determine the required financial support.** Any model would require government support, at least at the start, to cover investment and part of the operation cost. This is because the operation's sole income stems from selling electricity at a higher rate than the regular, which will not be enough to cover expenses. (For example, if a single fast charger is used by five EV customers per day with 10 kWh taken per charge, after 365 days 18,250 kWh of electricity has been sold. If this electricity is sold at a price double the regular price (at Nu 6 [US$0.1] per kWh instead of the normal price of Nu 3 (US$0.05) per kWh), the total income per fast charger is about US$912.50 per year, which is not enough to cover operations and maintenance costs.)
- **Identification of possible interest in the role of charging infrastructure operator.** Because knowledge about the EV market and RGoB plans for EVs are quite limited and the operator role is very new, a communication strategy may be needed to identify companies possibly interested in becoming the charging infrastructure operator.
- **Preparation of a service-level agreement with appropriate requirements.** Appropriate requirements from RGoB to the operator need to be prepared in a service-level agreement for a certain duration with indicators such as uptime and accessibility of the chargers, safety requirements, client access and satisfaction, cost efficiency, maximum selling price,

box continues next page

Box 6.13 First Assessment: Selecting a Charging Infrastructure Operator *(continued)*

marketing, and reporting. The agreement should also describe the allocation of risks, insurance, termination, and end of contract settlement.
- Development of a framework for future regulation of the market for EV charging and RGoB requirements for the charging operator. If it will be a commercial role, the contractual relationship and market organization must be decided. In practice, the first operator will have a monopoly at the start of the program.

Box 6.14 Cost Assumptions for Charging Equipment

The cost calculations for charging equipment are based on assumptions for the costs involved with home, public, workplace, and fast chargers, as summarized in table B6.14.1. Costs involved with adjustments to the BPC electricity distribution grid, increased energy losses, or aging of transformers because of EV developments have not been taken into account. Only the additional transformer capacity and cabling have been included in the calculations.

Table B6.14.1 Cost Assumptions for Charging Equipment

Charger	Price	Description
Home charger	US$200 (Nu 12,000)	• Mostly wall mounted on existing grid connection. • Price includes supply and installation of a circuit breaker and simple industrial socket.
Public slow chargers	US$1500 (Nu 90,000), with annual maintenance and operations cost of US$100 (Nu 6,000).	• Price is for a single charger connected to a new BPC connection. • Cost of signage and site preparation is not included.
Workplace charger	US$200 (Nu 12,000)	• Mostly wall mounted on existing grid connection.
Fast charger	US$30,000 (Nu 1,800,000), with yearly cost of US$2,000 (Nu 120,000) for maintenance and operations.	• Cost for fast charging is calculated by adding the costs of (a) hardware supply and ICT (US$15,000/Nu 900,000); (b) BPC grid connection and transformer (US$12,000/Nu 720,000) (average price, price dependent on cable distance); and (c) installation of chargers (US$3,000/Nu 180,000). • For the super high uptake scenario (scenario 3), cost is reduced by 30 percent because of the high number of chargers.

Costs and Financing Arrangements

Using the calculations of the estimated required number of chargers for each scenario ("Charging Infrastructure Requirements by Uptake Scenario and National Roll-Out," above, tables 6.5–6.7), a rough cost indication can be developed for the supply, installation, and operation of charging infrastructure. Assumptions about the costs of charging equipment are summarized in box 6.14; total estimated costs per scenario are shown in table 6.10. Estimated costs

Table 6.10 Indicative Cost per Scenario

Type	Low EV Uptake Scenario, in US$ (Nu)	High EV Uptake Scenario, in US$ (Nu)	Super High EV Uptake Scenario, US$ (Nu)
Home charging	60,000 (3,600,000)	206,000 (12,360,000)	859,000 (51,540,000)
Home public	240,000 (14,400,000)	817,000 (49,020,000)	3,403,000 (204,180,000)
Work	34,000 (2,040,000)	103,000 (6,180,000)	172,000 (10,320,000)
Fast charging	349,000 (20,940,000)	1,569,000 (94,140,000)	6,697,000 (401,820,000)
Total	684,000 (41,040,000)	2,696,000 (161,760,000)	11,131,000 (667,860,000)

Source: World Bank analysis.
Note: For the super high uptake scenario cost is reduced by 30 percent because of the high number of chargers. Other cost assumptions listed in box 6.14.

include only those for hardware; other costs such as those for organization and personnel, as well as costs involved with planning adjustments, capacity building, and the charging operator organization are not yet taken into account. Revenues are also not considered.

Although it serves only a small number of EV drivers, the cost for home public charging is a significant part of the total cost. This shows also the importance of home and work charging from a financial perspective.

The calculations also assume chargers will be technically operational until 2020. It is, however, possible that formal standards for normal and fast charging will be defined during this period. If many chargers have been installed from another type, these will have to be replaced before 2020, which means extra cost will be incurred.

Financing Arrangements

Globally, different approaches have been used for the financing and building of charging infrastructure. In Japan, for example, Nissan was the main investor in charging infrastructure, while in Estonia the federal government and Mitsubishi worked together to make those investments. And, although both these countries focus mainly on DC fast charging (using CHAdeMO), other countries such as the Netherlands focus more on AC private or public normal charging; this choice is very dependent on the demographic characteristics, road network, public investment, and strategic choices of regional and national governments.

Different arrangements may be used to pay for the various types of chargers.

- **Home chargers.** Home chargers can be supplied with the vehicle by the car manufacturer and financed by the customer. In many EV countries financial incentives by national or local governments are in place to support the installation of safe and intelligent chargers at home (Mock and Yang 2014)

- **Workplace chargers.** Workplace chargers can be installed and financed by the building owner or company renting the building, such as a government institution, hotel, or school. By providing free electricity some employers incentivize the use of EVs. If the driver has to pay for electricity, a payment system needs to be in place. This can generate income for the employer if the occupancy of the charger is high.

- **Public slow chargers.** The public slow chargers can be installed and regulated by the charging infrastructure operator (in the integrated model) or by a commercial party commissioned by the local government, such as Thimphu Thromde. The business model of selling electricity to recover cost has proved to be very difficult for public slow charging because of the high investment, low occupancy, and small premium on existing electricity price (Vollers 2013). It might be possible to have large companies like GE, Siemens, ABB, or Schneider supply part of this infrastructure.

- **Fast chargers.** The supply of fast chargers can partly be done by car manufacturers like Nissan, Tesla, and Mahindra to support EV sales. These stations, however, might support different charging standards and cannot be integrated in one asset management system of the charging infrastructure operator. This may lead to high maintenance and repair costs in the future. The model is also not very scalable because it is dependent on the willingness of the car manufacturers. Another option to pay for fast chargers, under an independent e-mobility model, is the investment by commercial parties in fast chargers on private ground (like petrol stations), with a long-term view on EV development. This business model, however, is dependent on a number of factors (electricity price, occupancy, and number of users) and has proved to be difficult in many countries; government support is always needed. While income can be generated from selling electricity at a premium, the price, however, cannot be too high compared to the normal electricity price to ensure high charger occupancy and keep EVs financially attractive. Monthly membership fees have also been used to recover some of the cost.

Grid Impact and Power Quality

The uptake of EVs will have an impact on the electricity grid, especially in the medium and long term when large numbers of vehicles enter the market. A government decision to incentivize EVs should thus have an effect on the long-term infrastructure planning of BPC, the public utility responsible for distributing electricity throughout the country and providing transmission access for generating stations for domestic supply as well as export.

Impacts on the Electricity Grid

The electricity system will be impacted through the following:

- **Increased electricity generation.** Electricity generation will increase as a result of the growth in the number of EVs; this increase will come on top of a growing use of electricity in general.

- **Increased need for grid renewal and upgrades.** The uptake of EVs may require an upgrading of the distribution grid because of the additional EV loads. This upgrade may be needed first in weaker areas of the grid with high demand (for example, with fast chargers or clusters of normal chargers). Parts of the grid may also need to be more frequently renewed as the lifetime of a transformer station reduces when loads increase.
- **Increased energy losses.** Energy losses in transport and transforming stations will increase as a result of the growing volume.
- **Changes in power quality.** The extra demand for electricity may lead to problems with the power quality, such as dips, swells, and harmonics. Drops in voltage are especially relevant in rural networks with long line length. With the use of single-phase chargers, an increase of unbalanced voltage can occur because of an uneven distribution of single-phased loads.
- **Need for improved IT systems.** EVs are a new source of power demand, and this demand can be influenced with intelligent communication between grid and vehicle. To be able to capture this possibility, IT systems for smart grids need to be in place.

Figure 6.7 illustrates why the grid may need an increased rate of upgrading. For each subnetwork (or network component such as a transformer), a margin between expected load and limit can be determined. If the ratio between expected annual peak load of a subnetwork and the peak load limit of the network is plotted, a curve such as that in figure 6.7 is shown. The networks that even without EVs (orange line) are already near the maximum will have to be upgraded earlier when EV loads are included (blue line). Box 6.15 provides a first assessment on the possible grid impact in Bhutan.

Global Experience with Charging Times and Peak Load EV

The effect of EV charging on the grid also depends on the time of charging in relation to the existing peak load on the grid. Global experience suggests that the two often coincide, with charging taking place during the existing peak in electricity consumption in the evening between 6 p.m. and 8 p.m. Also in Bhutan, the general peak load for energy consumption is in the evening (figure 6.8) at about 6 p.m. The size of the peak load (in MW) varies between winter and summer, in part because of the use of electric heating only in the winter (figure 6.9).

Charging times (and the amount charged) have been analyzed in various studies. Figure 6.10 for example illustrates the typical driving pattern of a driver in the Netherlands (using a nonspecific EV), showing both the arrival time and distance driven, with the latter indicating the required energy to recharge. The Z-axis indicates the number of events (out of 18,000 data points), showing that many people arrive around 6 p.m.

The charging behavior of EV drivers is detailed in the next two figures, showing charging at both a typical public charging station in the Netherlands (figure 6.11) and DC fast charging in the United States (figure 6.12). The graph for the Netherlands shows a peak for the public charging station on

Charging Infrastructure and Network Planning

Figure 6.7 Histogram of the Ratio between Expected Annual Peak Load of a Subnetwork and the Peak Load Limit of the Network

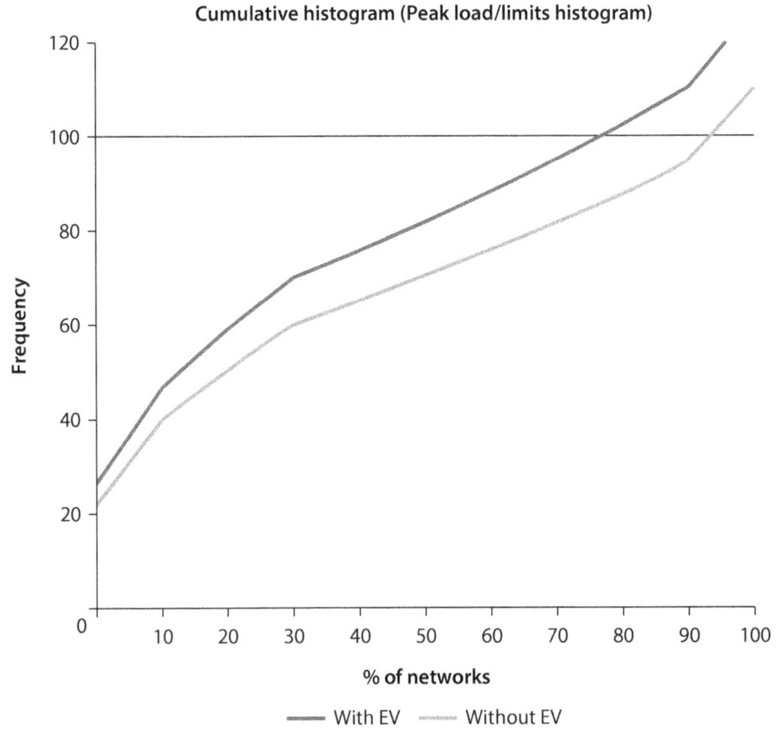

Cumulative histogram (Peak load/limits histogram)

[Chart: Frequency (y-axis, 0–120) vs % of networks (x-axis, 0–100), showing two curves: "With EV" and "Without EV"]

Source: Grid 4 vehicles 2011.
Note: EV = electric vehicle.

Box 6.15 First Assessment: Grid Impact

During the introduction phase of EVs, the possible impact on the grid—in terms of energy generation, grid renewal and upgrade, energy losses, power quality, and IT—should be investigated to prepare the grid for increased EV uptake in a later phase.

A practical way for BPC to gain experience with the impacts of EV charging would be the use of a testing fleet run by BPC itself. The fleet could include different vehicle types and brands, and also involve different charging setups at home and work. If equipped with data loggers, the vehicles could deliver valuable information on driving and charging behavior. Collaboration with the different OEMs is necessary to fully understand the integration of the charging process in the grid system.

Figure 6.8 Typical Morning and Evening Peak Load on the Electricity Grid in Bhutan

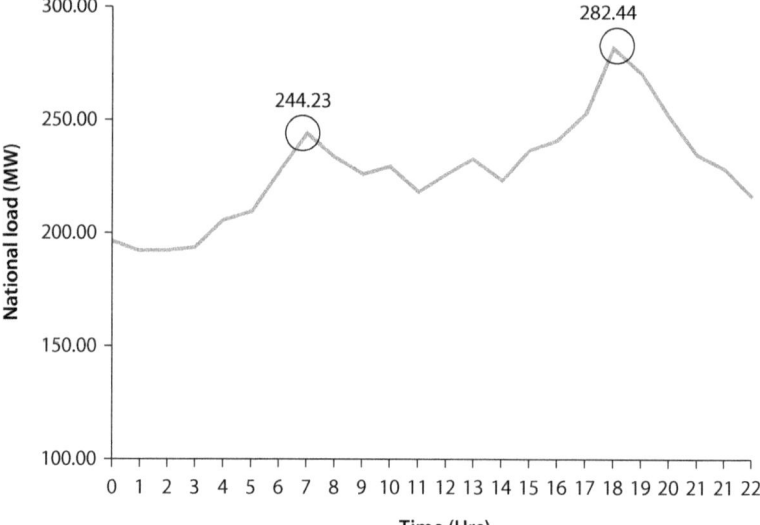

Source: Bhutan Power Corporation Limited, Ujjwal Dahal, February 25, 2014.

Figure 6.9 Daily Electricity Use Pattern (Load Profile), 2010

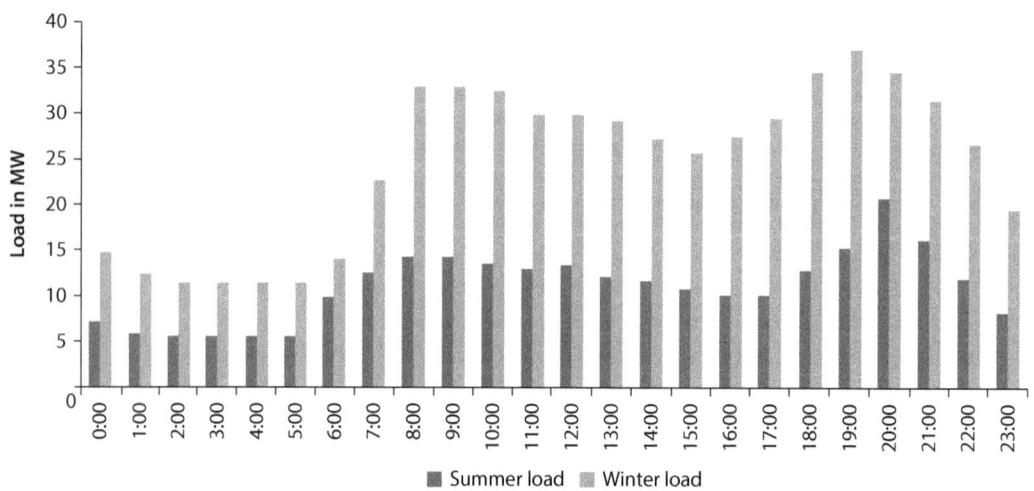

Source: Bhutan Power Corporation Limited 2012, Power Data Book.

Charging Infrastructure and Network Planning

Figure 6.10 Time and Distance Distribution for Average Travel

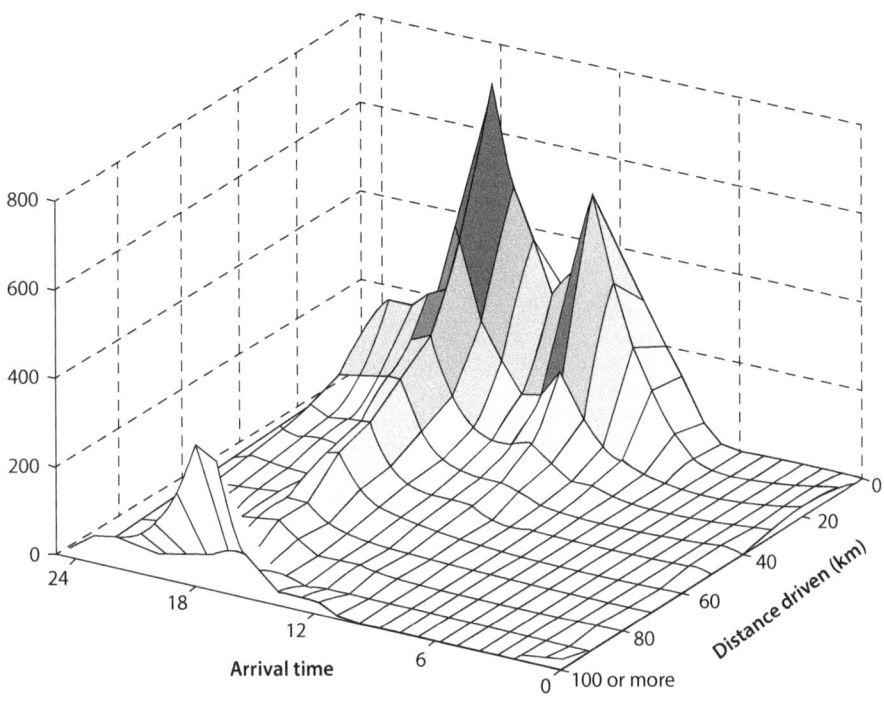

Source: Verzijlbergh et al. 2011. Used with the permission of Remco Verzijlbergh. Further permission required for reuse.

Figure 6.11 Typical Charging Profile at E-laad Public Charging Operator

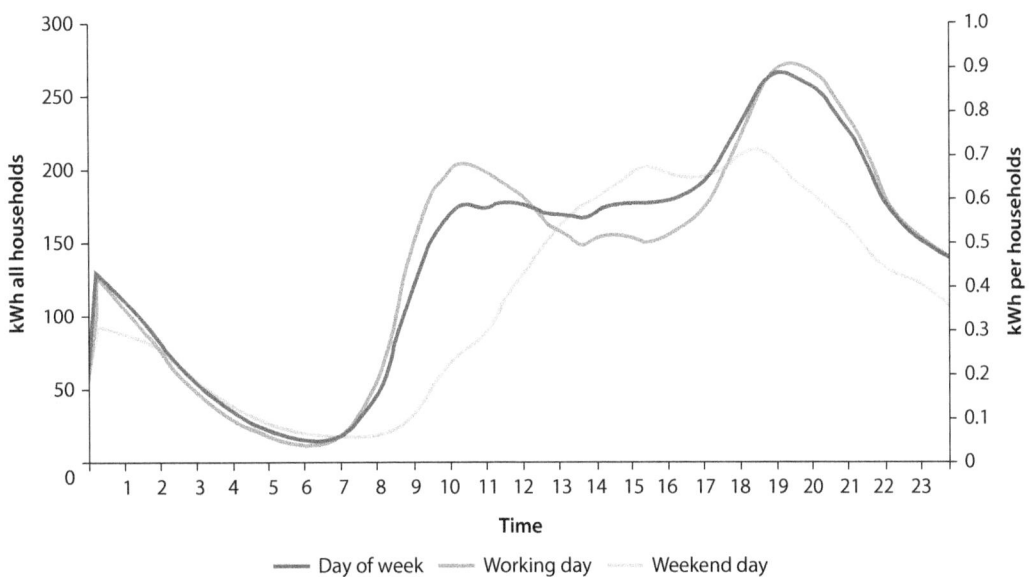

Source: E-laad, charging data from the Netherlands 2012.

The Bhutan Electric Vehicle Initiative • http://dx.doi.org/10.1596/978-1-4648-0741-1

Figure 6.12 DC Fast Charging Connection Time by Hour of the Day, by Region

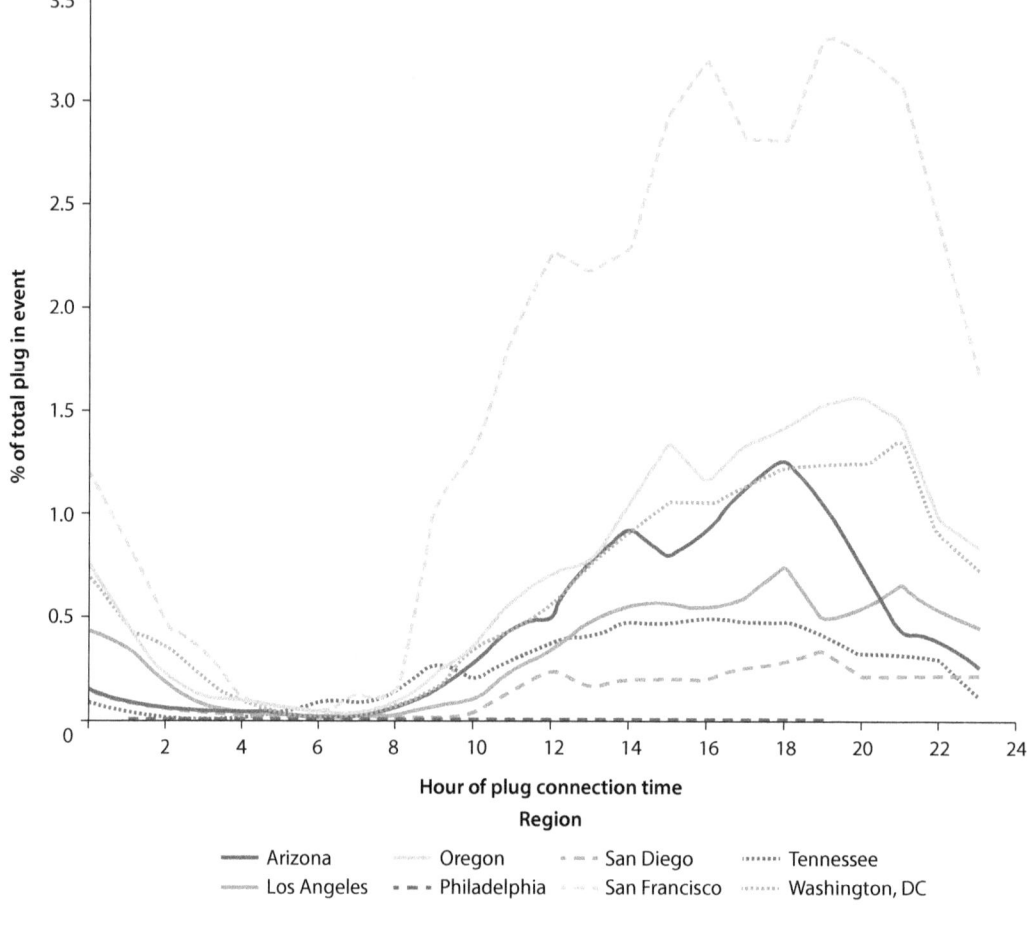

Source: EVProject United States, initial findings 2013.

weekdays between 8 a.m. and 11 a.m. and between 6 p.m. and 8 p.m. On the weekends, the peak falls between 3 p.m. and 7 p.m. In the United States, with the EVProject, the largest percentage of fast charging events is between 4 p.m. and 8 p.m.

The peaks for charging times in Bhutan and Thimphu are not yet known, and they may be different from the ones found in international studies because of smaller driving distances, different working hours, or different charging behavior. More information about these indicators will be needed to plan the charging network.

Impact by EV Uptake Scenario

In general, the impact on the electricity grid will depend on the uptake of EVs in Bhutan. Tables 6.11–6.13 show the estimated total energy use in 2020 and the maximum theoretical peak for the three scenarios of low, high, and super high

Table 6.11 Low EV Uptake Scenario: Total Energy Use and Maximum Peak Load in 2020

	Taxi	Government	Private	Total
Number of EVs in 2020	132	21	323	476
Energy use in 2020				
Average km per year	50,000	10,000	10,000	
Total energy use (GWh)	1.2	0.0	0.6	1.3
Charging peak (MW)	0.4	0.1	1.1	1.4
Percentage of total peak				0.4%

Source: World Bank analysis.
Note: Average kWh/km is assumed to be 0.19 kWh (based on actual user data); normal charging is at 3.3 kW. EV = electric vehicle; GWh = gigawatt-hour; MW = megawatt.

Table 6.12 High EV Uptake Scenario: Total Energy Use and Maximum Peak Load in 2020

	Taxi	Government	Private	Total
Number of EVs in 2020	441	62	969	1,472
Energy use in 2020				
Average km per year	50,000	10,000	10,000	
Total energy use (GWh)	4.1	0.1	1.8	6.0
Charging peak (MW)	1.5	0.2	3.2	4.9
Percentage of total peak				1.2%

Source: World Bank analysis.
Note: EV = electric vehicle; GWh = gigawatt-hour; MW = megawatt.

Table 6.13 Super High EV Uptake Scenario: Total Energy Use and Maximum Peak Load in 2020

	Taxi	Government	Private	Total
Number of EVs in 2020	4,414	104	1,615	6,132
Energy use in 2020				
Average km per year	50,000	10,000	10,000	
Total energy use (GWh)	41.2	0.2	3.0	44.4
Charging peak (MW)	14.6	0.3	5.3	20.2
Percentage of total peak				5.2%

Source: World Bank analysis.
Note: EV = electric vehicle; GWh = gigawatt-hour; MW = megawatt.

EV uptake in Bhutan. Calculations are based on the assumption that, on average, government and private vehicles drive 10,000 km per year and taxis drive 50,000 km per year. The peak use is calculated assuming *all* vehicles charge exactly at the same time at 3.3 kW (thus including normal charging and not fast charging). This total charging peak for EVs is compared to the total peak use of Bhutan in 2020, which is estimated at 390 MW (a 30 percent increase over the 300 MW peak use in 2014). For this calculation, other scenarios (such as all vehicles charging with 7 kW AC chargers or all 50 kW DC fast charging stations charging at the same time) have not been taken into consideration.

An assessment of the impact on the electricity grid should include an assessment of energy production, transmission, and distribution. See box 6.16 for a first assessment on the effects for energy use and the impact on the grid in Bhutan.

- **Energy production.** Over the next few years, energy production in Bhutan will increase significantly with the delivery of the country's hydropower projects, so the extra energy consumption for EVs will in general not have an impact. Difficulties, however, can arise in winter months: in 2014 hydropower production was not enough (260 MW) for the normal peak use (300 MW), and energy needed to be imported from India. This situation occurs during about three to five months of the year, only at peak hours. The EV charging, if done during peak, will in that case directly increase energy imports. Currently, imports are about 40 MW, depending on network conditions, generation scheduling, and other parameters; an additional EV load of 20 MW in the super high EV uptake scenario would generate a significant import increase. An energy banking system with India is under discussion, which may also need to include the EV loads in its load forecast model.
- **Energy transmission.** Transmission capacity will be increased with the additional power generation, and EV energy use can likely be absorbed because of its relatively small quantity. This, however, needs to be confirmed by BPC.
- **Energy distribution.** At the distribution level, EV charging can have an impact on the local level if, for example, many home chargers are installed in one area or if a hub with 10 fast chargers is installed.

Impact at Household Level

An average electric vehicle driving 10,000 km per year with 0.19 kWh/km efficiency consumes about 1,900 kWh per year. The average electricity use in Bhutan is about 800–3,000 kWh per household per annum (Palit and Garud 2010). If the vehicle is charged at home on the existing grid connection, the impact on household consumption is significant. In addition, the impact on the peak use at household level is significant, especially when a 7.7 kW charger

---

Box 6.16 First Assessment: Energy Use and Impact on the Grid

Total energy use and peak use are only a small percentage of the total peak use in 2020. However, during peak use hours in the winter months, the energy consumption by EVs has a direct negative impact. The impact on the grid needs to be investigated in further detail at the local distribution level to determine required investments. As the number of EVs increases, the need will grow to move charging to off-peak hours.

(an option on the Nissan Leaf) is installed. The calculated average power usage for an apartment is 15 kW,[4] which then is increased with 7.7 kW for the charger. This is a very significant increase of power usage by about 50 percent

The process for new buildings now includes a basic check by Thimphu Thromde on the electrical installation and a check by BPC before a new grid connection is installed. The check by Thimphu Thromde checks the expected electricity use per socket and a 60 percent demand factor. These procedures will have to be revised if a wide-scale EV program is implemented. In addition, the charger currently is a part of the electrical installation of the house and not included in the regulated assets base of BPC (after the meter). The safe installation of home chargers, however, is an important aspect that requires skilled technicians.

Fast Charging Locations and Grid Connection

The location for fast chargers needs to be selected in relation to the distance to the nearest 33/11 kV line (see also "EV Charging Options in Bhutan," above, and map 6.1). Following a first assessment, BPC reports the need for an additional transformer at each fast charger location.

A single 50 kW fast charging station requires a high-power grid connection (three-phase, 400V at 100A). When multiple outlets are used at maximum output at the same time, the grid connection needs to be a multitude of this. The underlying electricity grid needs to be able to deliver this at the required location.

Finally, the fast chargers also need a stable electricity supply without dips, swells, and harmonic events. An initial survey, performed by Tata Engineering (assigned by Nissan), indicated that the overall power quality in Bhutan is sufficient but that power quality is not very good at some specific locations. This has to be investigated further with BPC and the charging infrastructure operator.

Off-Peak Charging and Smart Charging

To limit the need for future investments in the electricity network, it is clear that EV charging will need to move to off-peak hours. Normally drivers would plug in their EVs after arriving at home, which is easy and comfortable, and charging may be needed for a later drive. To change this behavior, incentives need to be in place. Various methods can be used, such as increasing customer awareness, introducing variable tariff structures with a special low off-peak EV charging tariff, or (when a smart grid is in place) using an optimized smart charging process. Smart charging can reduce the impact on the grid, for example by starting the charging process when there is enough capacity on the grid (for example, at night or when there is a surplus of renewable energy) and demand is low. A simple solution could also be a small timer installed with the charging station to postpone the start of charging, or to program the car to finish charging by, for example, 6 a.m. The last two options, however, are dependent on a willingness to adjust behavior.

When grid-optimized smart charging is introduced, communication needs to be in place between the charging station, the EV, the electricity supplier, and the grid company (DSO), with the use of smart metering. One condition for introducing smart charging is the use of mode 3 charging with an intelligent charging station. Initial small-scale pilots with smart charging have been implemented around the world, and wide-scale implementation is expected at a later stage. Figure 6.13 illustrates the communication structure of smart EV charging.

Smart Grid Solutions and EV Innovation Projects

Several innovations help reduce the impact of EV charging on the grid. For example, EVs can be integrated in smart grid solutions with net balancing by EV batteries. The charging power and charging time will be adjusted to balance the electricity grid. In addition, locally produced energy from solar power can be used to charge EVs, with the charging moment optimized to make maximum use of the local production. Another innovation is the use of EVs as emergency power back-ups for households or buildings. These solutions are in an experimental stage and far from large-scale commercial implementation. A smart grid is often referred to as a system that includes a variety of operational and energy measures such as smart meters, renewable energy resources, and the possibility of energy storage. Figure 6.14 shows an example of a vehicle to home smart grid.

Figure 6.13 Communication Structure for Smart EV Charging

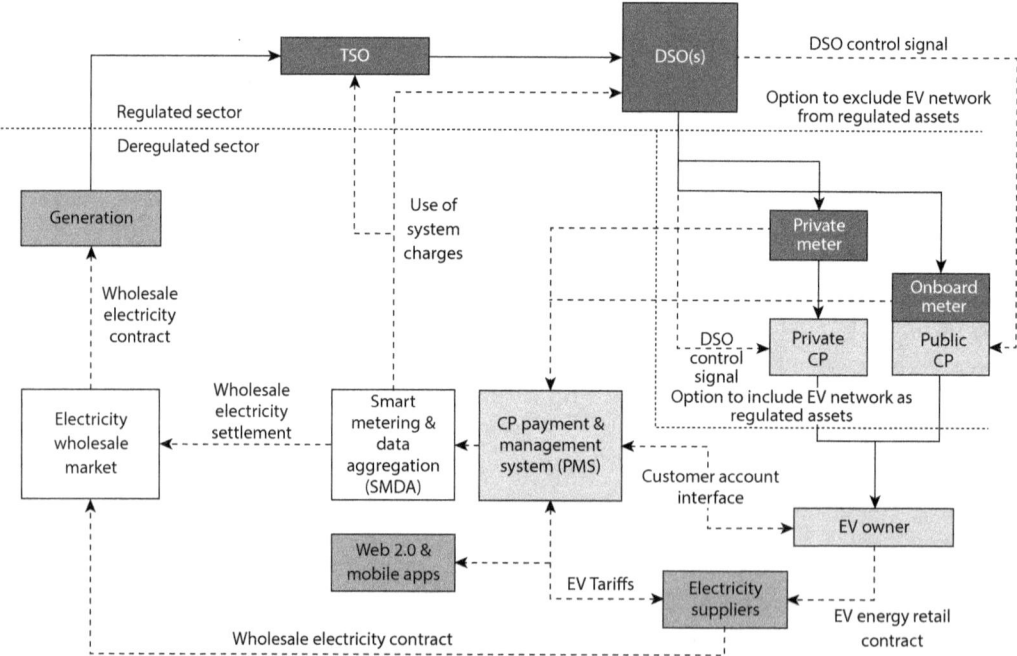

Source: World Bank.
Note: CP = charge point DSO = Distribution System Operator; EV = electric vehicle.

Figure 6.14 Vehicle-to-Home Smart Grid

Source: © Nissan. Used with permission. Further permission required for reuse.

Notes

1. In general, placement of charging infrastructure should be demand driven. Even chargers at strategically placed charging locations (for example train stations, village centers, and churches) without a specific group of users to use them will often be poorly used.
2. June 12, 2014: Tesla opens all patents of Supercharger technology and interest from Nissan (CHAdeMO) and BMW (Combo) to cooperate on joint development.
3. The bankruptcy of charging operator Ecotality (13,000 stations) in the United States despite a US$100 million (Nu 6 billion) loan from the Department of Energy is an example of the imbalance between investment and number of EV users.
4. Interview with representative of the Thimphu Thromde Electrical Engineering Department, May 2014.

References

Eurelectric. 2013. *Deploying Publicly Accessible Infrastructure for Electric Vehicles: How to Organize the Market?* Brussels, Belgium: Eurelectric.

FDT (Danish Association of Transport and Logistic Centres). 2013. *Danish Experiences in Setting Up Charging Infrastructure for Electric Vehicles with a Special Focus on Battery Swap Stations.* Aalborg, Denmark: FDT.

Grid for Vehicles. 2011. *Report with the Recommendations for the Grid Planning and Operation.*

ICCT (International Council on Clean Transportation). 2014. *Driving Electrification: A Global Comparison of Fiscal Incentive Policy for Electric Vehicles*. Washington, DC: ICCT.

Mock, P., and Z. Yang. 2014. *Driving Electrification. A Global Comparison of Fiscal Incentive Policy for Electric Vehicles*. Washington, DC: International Council on Clean Transportation.

Palit, Debajit, and Shirish Garud. 2010. "Energy Consumption in the Residential Sector in the Himalayan Kingdom of Bhutan." *Boiling Point* 58: 34–36.

Sierzchula, W., S. Bakker, K. Maat, and B. van Wee. 2014. "The Influence of Financial Incentives and Other Socioeconomic Factors on Electric Vehicle Adoption." *Energy Policy* 68: 183–194.

Trigg, T. and P. Telleen. 2013. *Global EV Outlook 2020*. IEA (International Energy Agency).

Verzijlbergh, R.A., Z. Lukszo, E. Veldman, J.G. Slootweg, and M. Ilic. 2011. "Deriving Electric Vehicle Charge Profiles from Driving Statistics." Paper prepared for the Power and Energy Society General Meeting, 2011 IEEE, San Diego, CA, July 24–29.

Vollers, Welmoed. 2013. "Sustainable Business Models for Public Charging Points." Graduate thesis. University of Technology, Eindhoven, Netherlands.

CHAPTER 7

Policy and Economic Analysis

Key Messages

- Based on the analysis, the current incentives program, which is sufficient to achieve the low electric vehicle (EV) uptake scenario, will not have a significant fiscal impact. (Fiscal impact from incentives for 2015–2020 is 0.30–0.45 percent of tax revenue and 0.20–0.21 percent of the national budget.)

- To achieve more ambitious targets under the high and super high EV uptake scenarios, fiscal impact will be more significant and a stronger incentives program will need to be put in place. (Fiscal impact is 0.99–1.51 percent of tax revenue and 0.68–0.72 percent of the national budget in the high uptake scenario and 3.60–5.47 percent of tax revenue and 2.45–2.61 percent of the national budget in the super high uptake scenario.)

- Impact on fuel imports will not be significant in the low and high EV uptake scenarios (annual avoided fuel imports in 2027 are, respectively, 1.57 percent and 5.1 percent of the total 2012 fuel imports for the low and high uptake scenarios) because diesel comprises the larger share of total fuel imports. In the super high EV uptake scenario, the annual avoided fuel imports in 2027 would be around 38 percent of total fuel imports in 2012. Impact, however, will be highly sensitive to changes in the global oil price.

- The net incremental impact on import will be mixed as the benefits of avoided fuel imports will be offset by incremental increases in import-intensive EVs and charging infrastructure. A net import reduction can be sustained only in the super high uptake scenario where the incremental impact will result in a net reduction of imports for 2018–2027, with a maximum reduction in impact of 1.81 percent in 2021 (based on import projections). Impact, however, will be highly sensitive to changes in the global oil price and inflation.

- The annual greenhouse gas (GHG) reduction benefits in 2027 are estimated to be 1.39 percent, 4.5 percent, and 33.6 percent of annual transport emission in 2000 for the low, high, and super high EV uptake scenarios, respectively. The total accumulated savings during 2015–2027 are estimated to be 17,276 tCO_2e (tons of carbon dioxide equivalent), 55,774 tCO_2e, and 416,897 tCO_2e for the three scenarios.

Overview of the Policy and Economic Analysis

Because the EV initiative in Bhutan has the objective to address environmental issues and reduce dependency on fossil fuel, a policy and economic analysis was carried out to assess the impact of the program in terms of fiscal impact, fuel imports, impact on the trade balance, and environmental benefits in terms of reduced GHG emissions. Required investments and spending for the initiative will stem from both the private and public sectors (see table 7.1), while the fiscal impact of the program is determined by the level of public investments to replace the government fleet with EVs, establish charging infrastructure, and provide an incentives program. The sections below present the results of the policy and economic analysis for the three different EV uptake scenarios.

EV Program Investment Requirements

Implementation of Bhutan's EV initiative will involve investments by both the private and the public sectors. Required investments will mainly involve the purchase of EVs and investments in charging infrastructure (roughly estimated to be about 10 percent of total investment needs; see also figure 7.1). As shown in table 7.2, during 2015–2020 the total investment requirement will range from Nu 728 million (US$12.13 million) in the low EV uptake scenario to Nu 7,793 million (US$129.88 million) in the super high uptake scenario (approximately 0.73–7.84 percent of 2012 gross domestic product [GDP]).

The next two figures illustrate how the total investment requirement is split between the investments in EV purchases and charging infrastructure (figure 7.1) and private and public contributions (figure 7.2). The EV program will be private led, with the private sector contributing about

Table 7.1 Overview of Private and Public Investments and Spending for the EV Initiative

Private investments and spending	Public investments and spending
• Purchase of EVs by private individuals and taxi drivers • Charging infrastructure at home and work	• Replacement of government fleet with EVs • Public charging infrastructure • Incentives program (lost tax revenue and cost subsidies)

Note: EV = electric vehicle.

Policy and Economic Analysis

Table 7.2 Investment Requirement of EV Program in Three EV Scenarios, 2015–2020

EV uptake scenario	Percentage of annual vehicle sales	Number of EVs per year	Total number of EVs in 2020	Total investment requirement 2015–2020 in million Nu (million US$)	Percentage of GDP
Low	1	79	476	728 (12.13)	0.73
High	3	245	1,472	2,362 (39.37)	2.38
Super high	5	1,022	6,132	7,793 (129.88)	7.84

Source: World Bank analysis.
Note: EV = electric vehicle; GDP = gross domestic product; Nu = Bhutanese ngultrum.

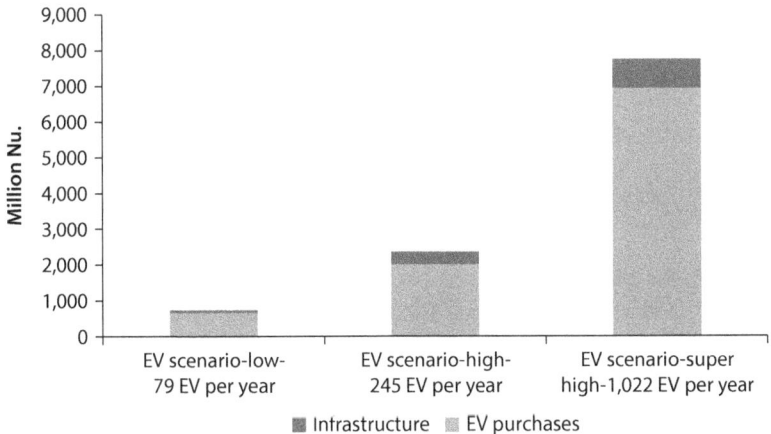

Figure 7.1 EV Purchases and Investment in Charging Infrastructure by Scenario, 2015–2020 (million Nu)

Source: World Bank analysis.
Note: EV = electric vehicle; Nu = Bhutanese ngultrum.

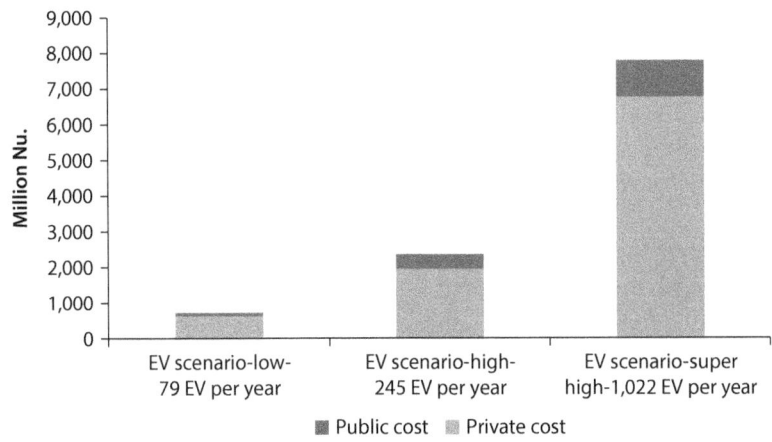

Figure 7.2 Private and Public Contribution by Scenario, 2015–2020 (million Nu)

Source: World Bank analysis.
Note: EV = electric vehicle; Nu = Bhutanese ngultrum.

The Bhutan Electric Vehicle Initiative • http://dx.doi.org/10.1596/978-1-4648-0741-1

87 percent of the total investment requirement. The public contribution for the remaining 13 percent will involve mainly the purchase of the government fleet and investment in charging infrastructure. In terms of vehicle purchases, based on the scenario assumptions, the segment of private vehicles will comprise the largest portion of the EV market, which means most of the EV purchases will be made by private individuals. On the contrary, for infrastructure investments, the majority (90 percent) of investment needs will be for public charging stations, although the private sector may need to pay for home and work charging.

Fiscal Impact

The fiscal impact of the EV program will stem from the public investments in government fleet replacement and public charging infrastructure, as well as the incentives program. The nature of the fiscal impact will depend on the design and level of incentives that will be used to promote EVs. To calculate the fiscal impact, the following assumptions were used for the fiscal incentives:

- Fiscal incentives will be provided throughout the EV program period (2015–2020).
- The incentive or cost subsidy is estimated only for the private vehicles segment; it is assumed that no additional fiscal incentives will be provided for the taxi segment.
- For the low EV uptake scenario, it is assumed that no additional incentive is used (beyond the current tax exemptions; see also table 5.2); fiscal impact is limited to uncollected taxes on EVs (45 percent sales tax and 45 percent customs duty of vehicles imported from countries other than India), representing a loss in revenue.
- For the high EV uptake scenario, it is assumed that a cost subsidy of 10 percent of the vehicle price (about Nu 126,000 or US$2,100 per vehicle) is provided; fiscal impact stems from the loss in revenue from uncollected taxes and the necessary public support to fund the subsidy.
- For the super high EV uptake scenario, it is assumed that a larger cost subsidy (35 percent of vehicle price or about Nu 220,000 or US$3,666) is provided to achieve a much more ambitious uptake target; fiscal impact will again result from uncollected taxes on EVs and public support to fund the subsidy.

Fiscal Impact from the Incentives Program

Table 7.3 shows the estimated total fiscal impact from the incentives under the three scenarios. As shown in the table, the current incentives program, sufficient for a low EV uptake target, will not have a significant fiscal impact. The total fiscal cost for incentives in this low uptake scenario is the amount of uncollected taxes on EVs, which is estimated to result in a loss in revenue of about Nu 593 million (US$9.88 million) during 2015–2020, equivalent to about 4 percent of total 2011/12 tax revenue.

Policy and Economic Analysis

Table 7.3 Estimated Total Fiscal Impact from Incentives by Scenario, 2015–2020

EV uptake scenario	Total uncollected taxes revenue million Nu (million US$)	Total cost subsidy million Nu (million US$)	Total fiscal impact from incentives million Nu (million US$)	Annual fiscal impact during 2015–2020 (as % of taxes revenue[a])	Annual fiscal impact during 2015–2020 (as % of budget[a])
Low	593 (9.9)	—	593 (9.9)	0.30–0.45	0.20–0.21
High	1,818 (30.3)	149 (2.5)	1,967 (32.8)	0.99–1.51	0.68–0.72
Super high	6,269 (104.5)	871 (14.5)	7,140 (119.0)	3.60–5.47	2.45–2.61

Source: World Bank analysis.
Note: EV = electric vehicle; Nu = Bhutanese ngultrum; — = not available.
a. Annual fiscal impact analysis is based on World Bank projections.

When achieving more ambitious targets under the high and super high uptake scenarios, fiscal impacts will be more significant and stronger incentives programs will be needed. In the high and super high EV uptake scenarios, the total fiscal impact from incentives is estimated at 13.4 percent and 42.7 percent of 2011/12 tax revenue. In terms of the annual fiscal impact during 2015–2020, the impact of losses in revenue will be higher in the earlier years and decrease gradually as tax revenue collection continues to improve. In the high uptake scenario, the annual fiscal impact will be between 0.99 and 1.51 percent of tax revenue and 0.68–0.72 percent of the national budget.

Fiscal Impact from Public Investments in Vehicles and Infrastructure

The additional budgetary support to invest in public infrastructure and replace the government fleet will also have fiscal impact. Table 7.4 shows the initial estimates of fiscal impact from public investment for the three EV uptake scenarios. Investment in charging infrastructure will be the largest component. This is, however, an initial estimate because the extent of fiscal support will depend on the financing and operating modality of the charging stations and who will pay for them.

Implications for Fiscal Sustainability

The broader fiscal implications of the EV program will largely depend on the actual EV uptake target and design of the EV program, particularly the

Table 7.4 Total Public Investment by Scenario, 2015–2020

EV uptake scenario	Total investment in EV purchases-government fleet, million Nu (million US$)	Total public investment in charging infrastructure, million Nu (million US$)	Total public investment, million Nu (million US$)	Percentage of 2011/2012 budget
Low	32 (0.53)	62 (1.03)	94 (1.57)	0.24
High	96 (1.60)	320 (5.33)	415 (6.92)	1.07
Super high	160 (2.67)	752 (12.53)	912 (15.20)	2.35

Source: World Bank analysis.
Note: Budget in FY 2011/2012 is Nu 38,843 million (US$647.38 million). EV = electric vehicle; Nu = Bhutanese ngultrum.

incentives and financing method. Several general trends and key issues are important when assessing the fiscal implications of an EV program for Bhutan:

- **Targeted level of uptake.** For the current incentives, fiscal sustainability is not a concern. Current incentives, for now, will not have a major impact on revenue collection, especially considering the new tax increases on vehicles and recent lifting of the import ban, which are expected to improve overall tax revenue and may offset the foregone revenue from EV tax exemptions. For more aggressive scenarios and higher levels of incentives, fiscal sustainability will be a key consideration.

- **Timing and method of financing.** The timing and method of program financing can have significant implications, especially in the case of a program with an ambitious uptake target and sizable fiscal impact. The method of financing— whether and how the private sector is involved—also plays a key role.

- **Fiscal situation.** Although the fiscal deficit is projected to remain low, in the short term the fiscal situation of Bhutan appears to be constrained with a projected decline in tax revenue collection as percentage of GDP in 2014/2015, combined with a continued tightening of expenditures given the persistent shortages of rupees and expected declines in foreign grants (World Bank 2014a). While in the medium to longer term, as a result of the completion of several hydropower projects, the fiscal situation is expected to improve significantly (with a boost in the domestic revenue-to-GDP ratio and broadly balanced fiscal deficit over the long term), the expected decline in external budgetary aid[1] as per capita income rises will necessitate fiscal consolidation to improve the macrofiscal framework. This implies that the government has limited fiscal space to implement a new initiative or capital investment by relying on its own revenue source. Additional public support to implement an EV initiative will have to rely on other sources of funding. This puts a constraint on any tax incentive program.

- **Competing needs.** Linked to Bhutan's fiscal situation is the consideration of competing needs for public spending. Competing needs and the greater emphasis on public spending efficiency also suggest limited fiscal space for capital investment in the near future (IMF 2014). External debt financing of the program, however, may be available in the long term, after country-specific indicative thresholds improve following the period of rapid hydropower development.

- **Recurring fiscal implications.** Any recurring fiscal burden should be carefully planned for. Based on international experience, public support is needed to subsidize the capital investment and operation of public charging infrastructure. The analysis, however, has not taken into consideration any recurring fiscal implications from public charging infrastructure. The nature and extent of

Policy and Economic Analysis

fiscal impact will depend on the financial modality of investing and operating charging stations and how the cost will be split between public and private, which cannot currently be assessed. The financing modality should factor in any recurring costs required over the project life from the beginning, who will pay, if public support is needed, and whether it will be affordable from a fiscal perspective.

- **Review and adjustment of incentives.** In the medium term, as grants financing is expected to decline and the government needs to make more space for both recurrent and capital spending, the current tax exemptions should be subject to review to better assess impact on total revenue and fiscal position, especially if the government decides to pursue a more ambitious EV uptake target that could lead to more significant impact on revenue.

Fuel Import Benefits

Overview of the Fuel Import Situation

Fuel imports have been growing rapidly over the last ten years and recently at a rate much faster than total imports (figure 7.3); fuel imports on average grew 25 percent per annum during 2003–2012, while total imports grew at 10 percent per annum in the same period. The number of vehicles in the country increased by 11 percent over this period. Fuel imports constitute an important part of total imports, with diesel imports ranking first, accounting for 6.8 percent of total imports in terms of value, and gasoline imports ranking sixth with a share of 2.5 percent in 2011 (figure 7.4). Other items in the top ten of imports are

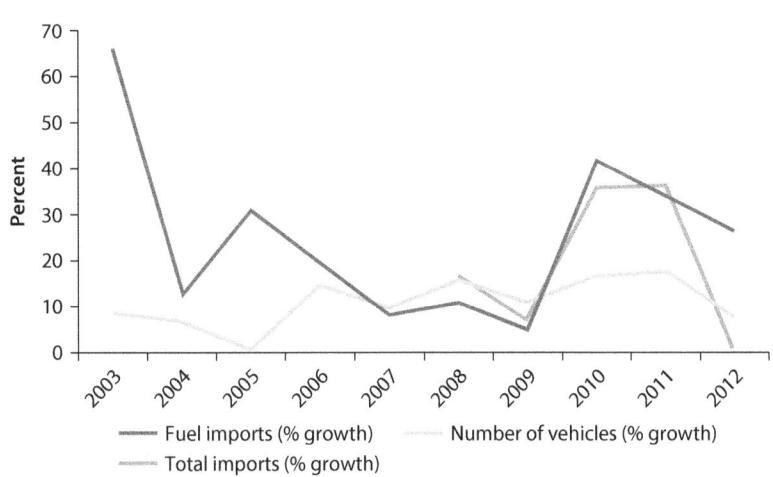

Figure 7.3 Percentage Growth in Number of Vehicles, Fuel Imports, and Total Imports, 2003–2012

Source: National Statistics Bureau.

Figure 7.4 Fuel and Vehicle Imports in 2011

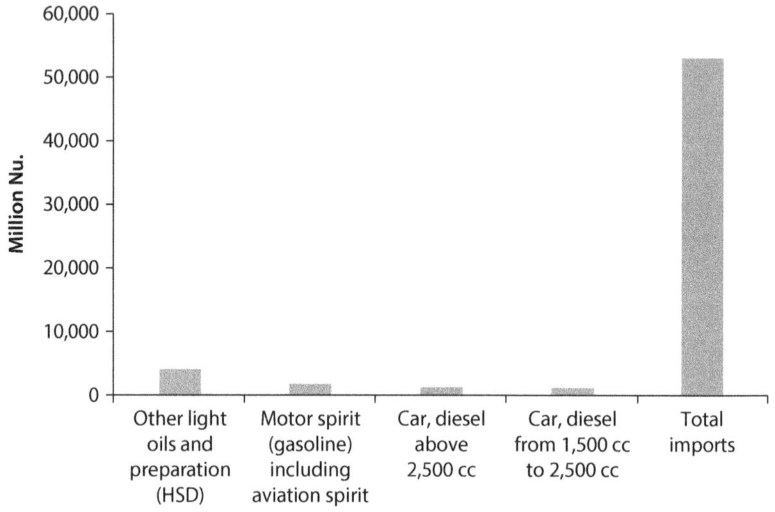

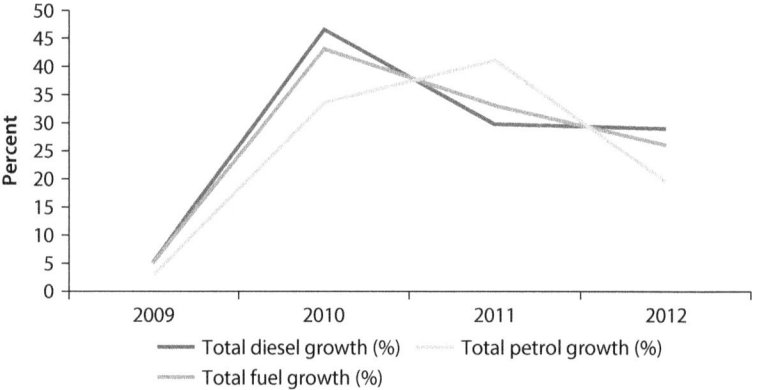

Source: Department of Revenue and Customs, Ministry of Finance.
Note: "Car, diesel" refers to high speed diesel (HSD).

Figure 7.5 Fuel, Diesel, and Petro Imports by Value, 2009–2012 (% Growth)

Source: National Statistics Bureau.
Note: Nu = Bhutanese ngultrum.

various capital goods (metals, machinery, and equipment), together accounting for about 33 percent of total imports in 2012.

Fuel imports are driven by the demand for both diesel and petrol, but demand for diesel is growing at a slightly faster rate that is driven largely by rapid hydropower development. During 2008–2012, the growth rate for fuel imports was 27 percent by value, while growth rates for diesel and petrol were 28 percent and 24 percent, respectively (figure 7.5). However, as shown in the figures, growth in

Policy and Economic Analysis

Figure 7.6 Fuel, Diesel, and Petro Imports by Volume, 2009–2012 (% Growth)

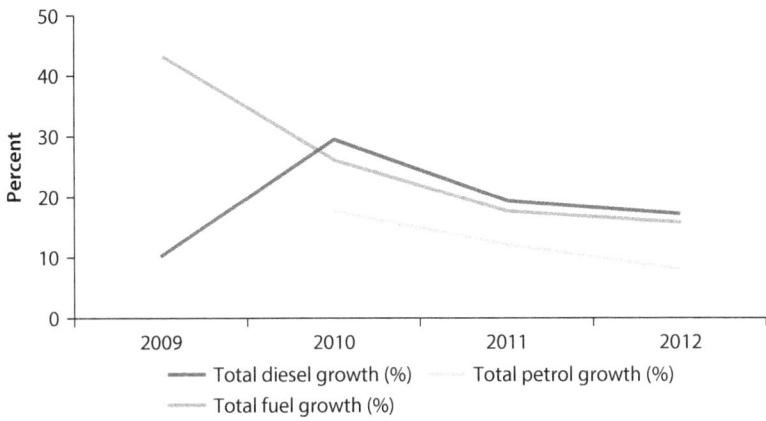

Source: National Statistics Bureau.

fuel imports was driven more by real demand (volume growth) than price increases. In the last three years, the volume of diesel imports grew by 66 percent, while the volume of petrol imports increased by 38 percent and total fuel imports volume by 60 percent (figure 7.6). In July 2014, the government for the first time imposed a fuel tax of 5 percent.

EV Program Impact on Fuel Imports

The three scenarios for EV uptake all have a different impact on fuel imports; with each EV replacement the country will benefit from avoided fuel import for the amount of fuel that an internal combustion engine (ICE) vehicle would have consumed. In the high EV uptake scenario, the annual avoided fuel imports in 2027 will be about 5.1 percent of total 2012 fuel imports (or 20 percent of 2012 gasoline imports). For the low uptake scenario, this number will be about 1.57 percent of total fuel imports or 6 percent of gasoline imports in 2012, while in the super high uptake scenario the annual avoided fuel imports in 2027 will be about 38 percent of total fuel imports in 2012 or 147 percent of 2012 gasoline imports.

Figure 7.7 shows the impact of different EV scenarios in terms of avoided fuel imports. As illustrated in the figure, the avoided fuel imports will gradually increase in value with the rising number of EVs in the country and an escalating fuel price, reaching their highest potential in 2027. According to the analysis, most of the avoided fuel imports will come from the EVs in the taxi segment.

For the analysis, it was assumed that the EV market will be developed sustainably in the long run and that EVs will be replaced by EVs. The analysis thus covers the period from 2015 until 2027 to include the EV program implementation period (2015–2020) and match the assumed life span of an EV (eight years). The calculation then assumed that the number of EVs in the fleet achieved in the year 2020 will continue at least until 2027 even without an incentives program.

The Bhutan Electric Vehicle Initiative • http://dx.doi.org/10.1596/978-1-4648-0741-1

Figure 7.7 Avoided Fuel Import by Scenario, 2015–2027

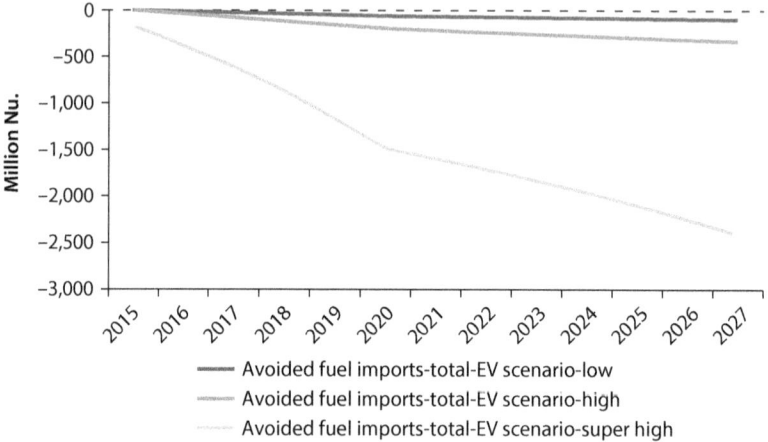

Source: World Bank analysis.
Note: EV = electric vehicle; Nu = Bhutanese ngultrum.

For the calculations, this means that, for an EV purchased in year one (2015) and disposed of at the end of year eight (2022), it is assumed that this EV will be replaced by a new EV in year nine (2023), leaving the number of EVs constant. This assumption corresponds to the objective of an EV program that aims to enable the market, allowing the government to phase out incentives once EVs can successfully penetrate the market on a commercial basis.

Impact on Trade Balance

The EV initiative and resulting increase in EVs in Bhutan will have a mixed impact on imports. On the one hand, investment in the initiative is import intensive because the vehicles and a large part of the charging infrastructure will be imported. On the other hand, avoided fuel consumption will reduce the demand for fuel imports in the longer run and have a positive impact on the trade balance.

Current Trade Balance and Import Situation

Bhutan's persistent trade deficit is a key macroeconomic issue for the country, and the EV program is expected to help contribute to curbing imports and improving the trade balance. Figure 7.8 illustrates the country's balance of payments, showing that it runs a large and growing current account deficit that currently stands at about 10 percent of GDP. The trade deficit has been growing in the last three years, along with an increase in imports (table 7.5). The current account deficit is projected to deteriorate over the medium term because of the strong growth in import demand associated with the construction phase of the hydropower projects. From 2020 onward, however, the situation is projected

Policy and Economic Analysis

Figure 7.8 Bhutan's Balance of Payments

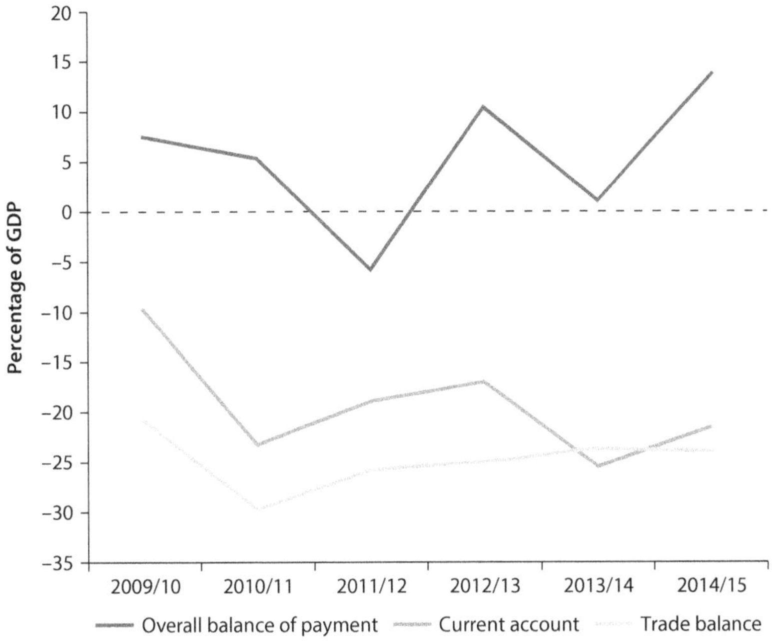

[Line chart showing Percentage of GDP from 2009/10 to 2014/15 for Overall balance of payment, Current account, and Trade balance]

Source: Royal Monetary Authority, graph from World Bank 2014a.

Table 7.5 External Trade in Million Nu

Trade	2009/10	2010/11	2011/12
Exports	25,401.8	30,160.1	29,890.4
Imports	−39,339.9	−53,705.0	−53,875.5
Balance of trade	−13,938.2	−23,544.9	−23,985.1

Source: National Statistics Bureau.

to improve when electricity exports are likely to more than triple compared to current levels, leading to a decline in the current account deficit and balance of payment surpluses (IMF 2014).

Total Imports under the EV Program

According to the analysis, a high EV uptake scenario will result in total imports during 2015–2020 of Nu 2,278 million (US$37.97 million), which is equivalent to 4.2 percent of total imports in 2012. Estimates of EV related imports for the three scenarios for 2015–2020 are shown in table 7.6.

Net Incremental Impact on Imports

To assess the net impact from the EV program and policy on imports over the study period, an incremental analysis was used. Implementation of the EV program will lead to three major incremental impacts on imports stemming from

Table 7.6 Total Impact of Imports from the EV Program, 2015–2020 (Million Nu and Percentage of 2012 Imports)

EV uptake scenario	Low	High	Super high
Total imports for electric vehicles and charging infrastructure 2015–2020 (%)	717 (1.3%)	2,278 (4.2%)	7,580 (14.1%)
Electric vehicle imports 2015–2020 (%)	659 (1%)	2,020 (4%)	6,966 (13%)
Charging infrastructure imports 2015–2020 (%)	58 (0.1%)	258 (0.5%)	614 (1.1%)

Source: World Bank analysis.
Note: Percentage expressed as percentage of 2012 imports (Nu 53,875.5).

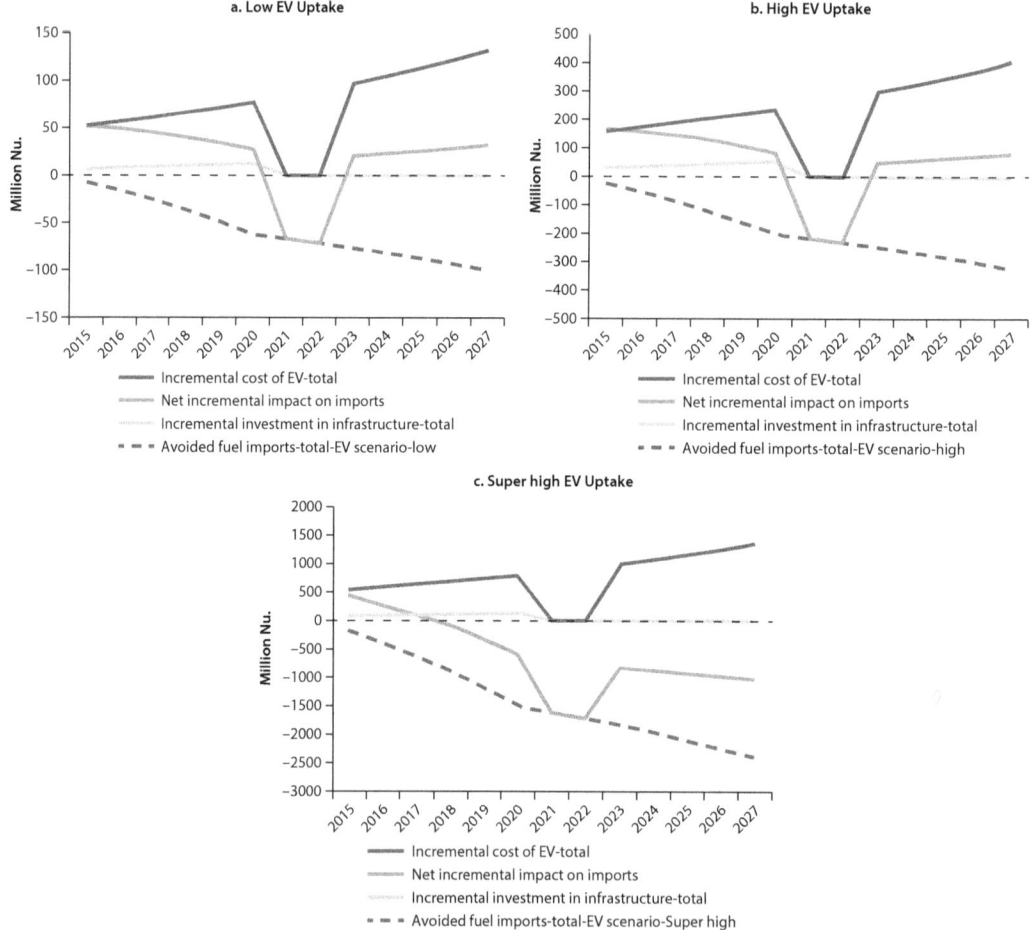

Figure 7.9 Incremental Impact on Imports, 2015–2027

Source: World Bank analysis.

(a) the import of EVs (because EVs are more expensive than ICE vehicles), (b) the installation of charging infrastructure, and (c) avoided fuel imports. Figure 7.9 illustrates the results for the three EV uptake scenarios.

Overall, as illustrated in the figures, the net incremental impact is first characterized by a short-term net increase in imports due to the higher prices of EVs (compared to ICE vehicles) and the cost of the imported charging infrastructure. After this initial investment period, the net impact starts to be negative. Because it is assumed that the number of EVs on the street will remain constant after 2020 (with EVs continually replaced by other EVs), imports will rise again in 2023 as people replace their eight-year-old EVs. Meanwhile, the fuel impact will continue to grow based on the assumption of annual fuel price increases. If the impact of the avoided fuel imports is large enough, such as in the super high scenario, it will result in a negative net impact on imports in the later years even with recurring imports from vehicles replacement.

In the low and high EV uptake scenarios, the net import reduction as a result of the EV program will be achieved only in the two years without replacements in the EV fleet; in those years the annual net reduction will be 0.07 percent and 0.24 percent, respectively. The net import reduction can be sustained only in the super high uptake scenario, where the incremental impact will result in a net reduction on imports in 2018 and continue until 2027 with a maximum impact of 1.81 percent reduction in 2021 (based on import projections). Figures 7.10 and 7.11 show the net impact of the three scenarios in value and as a share of imports. The net incremental impact on imports, however, is highly sensitive to the assumptions on future fuel price increases and inflation.

Net incremental impact on imports for different segments—private vehicles, taxis, and government fleet—under the high uptake scenario is shown in figure 7.12. The impact of the taxi segment is the largest and dominates the savings profile for the entire scenario. For taxis, the net incremental impact is

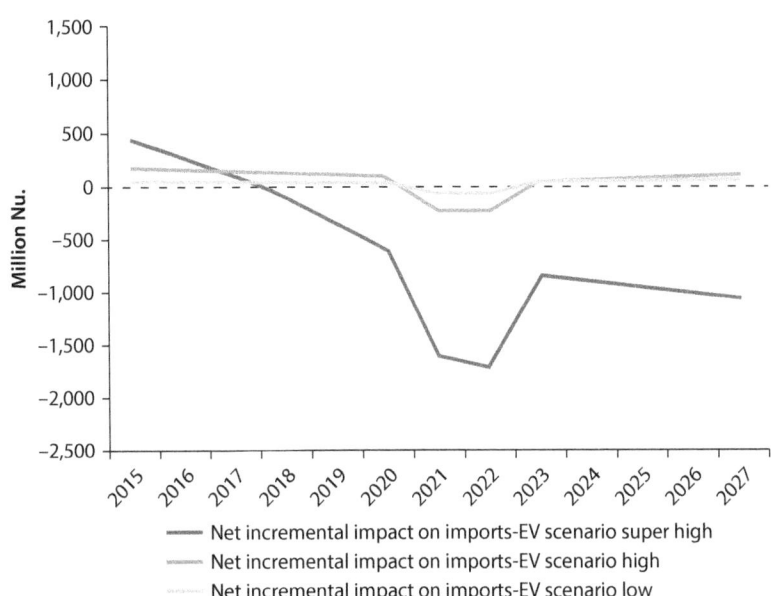

Figure 7.10 Net Incremental Impact on Imports by Scenario in Value

Source: World Bank analysis.

Figure 7.11 Net Incremental Impact on Imports by Scenario as Percentage of Future Imports Projection

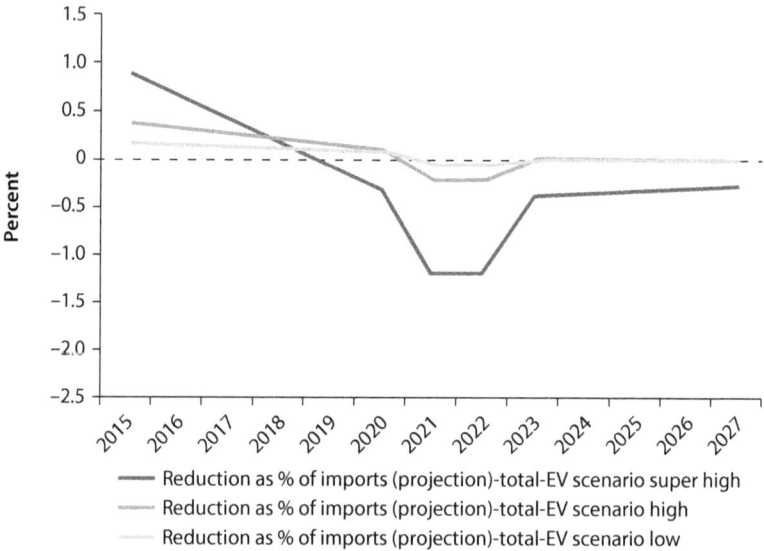

Source: World Bank analysis.

Figure 7.12 Net Incremental Impact on Imports for Private Vehicles, Taxis, and Government Fleet, High EV Uptake Scenario

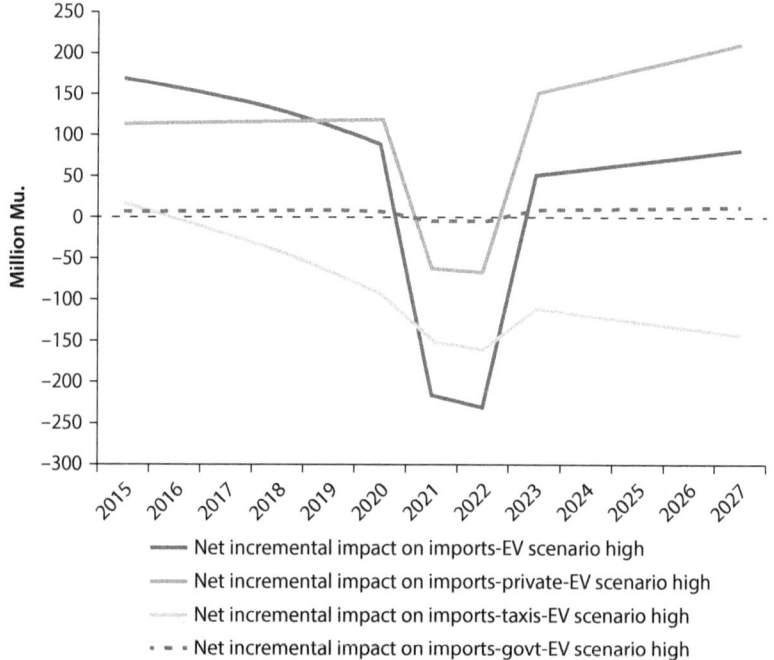

Source: World Bank analysis.

negative earlier than for the private and government segments. The taxi segment also contributes to the net reduction in imports in later years when vehicle imports rise again as a result of vehicle replacements, even if the net impact for private and government fleet is positive (a net increase in imports).

Policy implications from the net incremental impact analysis are twofold. First, the analysis makes clear that future increases in the fuel price and the price of EVs will largely determine whether the net impact on import will be positive or negative in the longer run. Second, the results show that a more gradual or phased approach to target setting will help smooth out the impact on imports. As the net incremental impact on imports is characterized by short-term costs (during the initial investment period) and long-term benefits (the savings from avoided fuel occur more gradually), the EV policy for the medium term should take into consideration the broader trend of the external trade situation to ensure minimal impact by the program on the trade balance and currency shortage. The extent and profile of the impact on import will also largely depend on how the target is set and on the implementation. A gradual approach to EV market development will allow for a moderate impact on the trade balance in the short run and sustained impact from fuel import reduction in the long run.

Environmental Benefits

The EV initiative has two key environmental benefits: (a) reducing air pollution and (b) reducing GHG emissions, thus contributing to climate change mitigation.

Air Pollution

The EV initiative at a significant penetration rate can have a large effect on urban air pollution because poor fuel quality and increased traffic are both major sources of urban air pollution in the Thimphu area. In fact, urban air pollution is one of the key environmental problems in Thimphu, and the dust from construction and road works and transport-related pollution are the two major sources of pollution. The vehicle-based air pollution is caused by poor fuel quality, inefficient combustion of fuel, and increased traffic around Thimphu Valley (World Bank 2014b). However, because no systematic data are collected on air quality due to lack of measuring devices, it is difficult to assess the level of impact of the EV initiative on the overall air pollution problem. As a result, the impact of the EV initiative on air pollution is not assessed in detail in this report.

GHG Emission Reduction

Bhutan is a net sink for GHG emissions. According to the Second National Communication in 2000 (National Environment Commission 2011), total GHG emissions excluding land use change and forestry (LUCF) were 1.559 million tons of CO_2-equivalent (tCO_2e) in 2000; total emissions composed of emissions related to energy (0.27 million tCO_2e), industrial processes (0.238 million tCO_2e), agriculture (1.005 million tCO_2e), and waste (0.046 million tCO_2e).

With a CO_2 sequestration from forestry and land use in 2000 amounting to 6.302 million tCO_2, total GHG emissions including LUCF are estimated to be –4.750 million tCO_2e, which indicates that Bhutan is a net sink. The transport sector emitted 0.118 million tCO_2e from fuel combustion activities, accounting for about 45 percent of all energy-related emissions or 7.5 percent of national GHG emissions.

Table 7.7 and figure 7.13 illustrate annual and total accumulated amounts for avoided GHG emissions by 2027. In the high uptake scenario, the total accumulated savings by 2027 are 55,774 tCO_2e, while the annual avoided GHG emissions in 2027 are 5,312 tCO_2e, representing about 4.5 percent of the annual transport emissions in 2000. Similar to other impacts, taxis contribute substantially to the overall savings profile (figure 7.14).

Table 7.7 Avoided GHG Emissions by Scenario (tCO_2e)

EV uptake scenario	Low	High	Super high
Annual GHG savings in 2027	1,645	5,312	39,705
Percentage of annual transport emission in 2000	1.39	4.5	33.6
Accumulated GHG savings 2015–2027	17,276	55,774	416,897

Source: World Bank analysis.
Note: GHG = greenhouse gas; tCO_2e = tons of carbon dioxide equivalent.

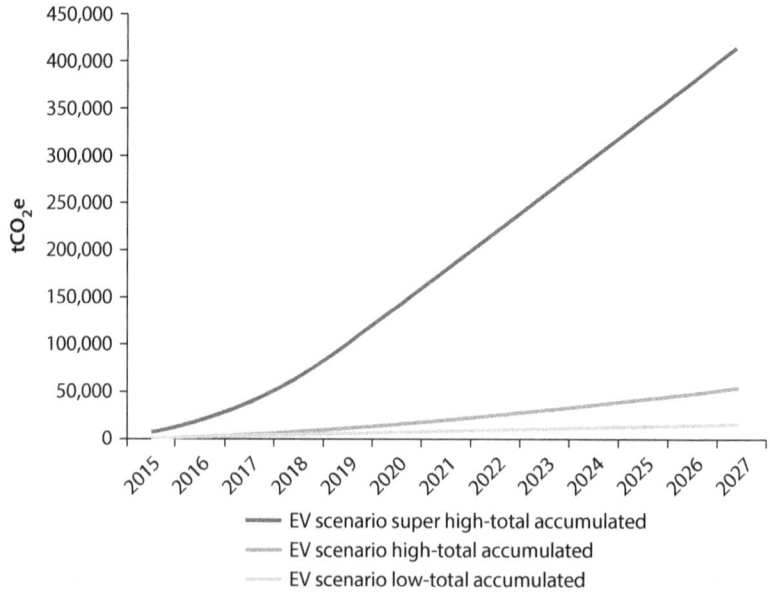

Figure 7.13 Accumulated Avoided GHG Emissions by Scenario, 2015–2027

Source: World Bank analysis.
Note: EV = electric vehicle; GHG = greenhouse gas; tCO_2e = tons of carbon dioxide equivalent.

Figure 7.14 Accumulated Avoided GHG Emission from Private Vehicles, Taxis, and Government Fleet in High EV Uptake Scenario, 2015–2027

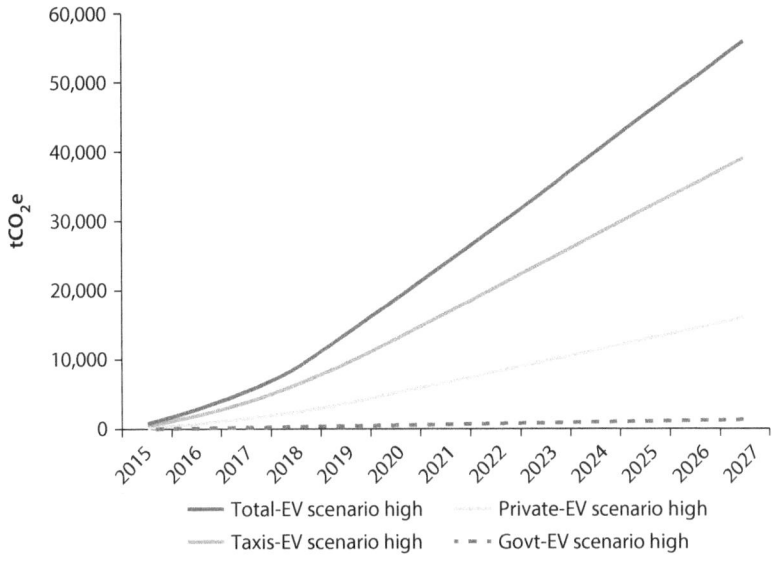

Source: World Bank analysis.
Note: EV = electric vehicle; GHG = greenhouse gas; tCO_2e = tons of carbon dioxide equivalent.

Table 7.8 Social Values of Avoided GHG Emissions in Different EV Scenarios

EV uptake scenario	Low	High	Super high
Social value of annual avoided GHG emissions in 2027 in million U.S. dollars (million Nu)	0.05 (3)	0.16 (9.6)	1.2 (72)
Accumulated social value of avoided GHG emissions 2015–2027 in million U.S. dollars (million Nu)	0.52 (31.2)	1.67 (100.2)	12.5 (750)

Source: World Bank analysis.
Note: EV = electric vehicle; GHG = greenhouse gas; Nu = Bhutanese ngultrum.

Reducing GHG emissions brings external benefits to society, as well as global benefits. Although the exact value of reduced GHG emissions (or costs from increased emission) from a society's perspective is still widely debated, the social value of avoided GHG emission in the different scenarios is estimated using the World Bank's recently published Guidance Note on Social Value of Carbon in Project Appraisal in 2014 (World Bank 2014c). At a social value of US$30 (Nu 1,800) per one ton of carbon emissions in 2015, the total accumulated social value of avoided GHG emissions is estimated to be US$0.52 million (Nu 31.2 million), US$1.67 million (Nu 100.2 million), and US$12.5 million (Nu 750 million) in the low, high, and super high uptake scenarios, respectively (table 7.8). The annual social value of avoided GHG emissions at the end of the EV program in 2027 is estimated to be US$0.05 million, US$0.16 million, and US$1.2 million in the low, high, and super high uptake scenarios.

Note

1. Today, domestic revenue covers about 67 percent of total expenditures, with 52 percent of current expenditures and 31 percent of capital expenditures funded by domestic resources and the rest by external grants and loans.

References

IMF (International Monetary Fund). 2014. *Bhutan Debt Sustainability Analysis.* Washington, DC: IMF.

National Environment Commission. 2011. *Second National Communication to the UNFCC.* Thimphu: Royal Government of Bhutan.

World Bank. 2014a. *Bhutan—Development Update.* Washington, DC: World Bank.

———. 2014b. *Note on Green Growth.* Washington, DC: World Bank.

———. 2014c. *Investment Project Financing Economic Analysis Guidance Note (for internal use by Bank staff).* Washington, DC: World Bank.

CHAPTER 8

Stakeholders and Public Transport

Key Messages

- The electric vehicle (EV) initiative will involve a diverse group of stakeholders in the public and private sectors whose interests should be considered by EV policy makers.
- Because of the high cost of EVs, only the higher- and upper-middle-income households will be target beneficiaries of the program. However, EVs will also generate public benefits in terms of reduced fuel imports, improved urban air pollution, and reduced greenhouse gas (GHG) emissions.
- Public transport is a green transport policy that has broader coverage of the urban population and can also achieve environmental benefits; in most countries, governments use EV initiatives as part of a broader green transport program.

Stakeholder Analysis

Stakeholder Mapping

The EV initiative will involve various stakeholders, who will all play different roles in the planning and implementation of the EV policy. How stakeholders are engaged in the policy will be one of the key factors for successful implementation.

In general, two main groups of stakeholders can be distinguished. The first group involves those who will directly participate in the EV market development, such as EV consumers, EV dealerships, parts and maintenance businesses, the power company, ministries and government agencies in charge of EV policies (for example financial, transport, climate, and communication), Thimphu and other municipalities, taxi associations, charging station operators, and financial institutions providing loans to EV consumers. For these stakeholders, the key policy implication is ensuring strong stakeholder participation, including building understanding and capacity, and enhancing cooperation and collaboration to move the EV program to the next phase. International experiences with this are discussed later in this section. The second group of stakeholders is formed by a broader group of actors who will be impacted directly or indirectly by the

EV policy. In addition to the first group, this group includes related business and industries that are competitors to EV (such as gas stations and dealers for internal combustion engine [ICE] vehicles), buyers of ICE vehicles, non-EV taxi drivers, and secondhand car markets.

Stakeholders Assessment

An initial stakeholder assessment suggested that the groups that are likely to benefit from the initiative are EV buyers, EV-related industries and businesses (including EV dealerships), charging station operators (as a new business), importers of related parts and equipment, and the power company (through increased sales of electricity). In addition, the secondhand car market may also benefit from the latest tax increases if consumers find they cannot afford a new car and decide to buy a secondhand one instead. The general public will benefit from the reductions in air pollution, CO_2 emissions, and noise compared to the use of ICE cars.

In contrast, other stakeholders will be negatively affected. Buyers of new ICE vehicles, for example, will be worse off as they will have to pay higher taxes for an ICE vehicle. ICE dealerships and fuel station owners also have the potential to lose some market share to the EV market. Some public programs may positively benefit from the increased revenue from the green tax, but others may be negatively affected by the foregone tax revenue as a result of the tax exemptions for EV, which could have been used for other public services or investment. Table 8.1 summarizes the results of the stakeholder assessment.

Table 8.1 Initial Assessment of Stakeholder Impact

Groups better-off	Gains	Groups worse off	Losses
EV buyers	Benefit from fiscal incentive (i.e., tax exemption)	Buyers of new ICE cars	Reduced affordability from higher taxes
EV dealerships	Increases in business opportunities	ICE dealerships	Potential reduced sales from higher penetration of EV market share (e.g., dealership)
Power/grid companies	Increases in business opportunities	Oil company	Reduced fuel sales as EV fleet grows larger
Other EV-related business and industries, e.g., charging operators, repair and maintenance, spare parts	Increases in business opportunities	Public programs	Forgone tax revenue from EVs could have been used for other public investment programs targeted at the poor
Public programs	Increased tax revenue from Green Tax to support public programs		
General public	Reduced local pollutant and CO_2 emission, reduced noise from ICE vehicles		
Secondhand market ICE vehicles	Higher taxes increase prices of new ICE cars and make secondhand market more competitive		

Source: World Bank analysis.
Note: EV = electric vehicle; ICE = internal combustion engine.

A phased approach to EV development will help all stakeholders adjust to the emergence of this new market in the auto sector. For stakeholders that will participate in the new market, capacity building, awareness raising, and knowledge sharing will be key for steadily building the EV sector in Bhutan. Similarly, for stakeholders that may be negatively impacted, a gradual approach to the development of the EV market will help players adjust to potential impact.

EV Beneficiaries

The EV initiative will primarily target private vehicle owners, making them the major beneficiaries. Among them, EV owners residing in the urban area and especially in Thimphu Dzongkhag are likely to benefit the most as EV characteristics are more suitable for urban use. The benefits for this group will be decreases in monthly fuel expenses, decreases in total cost of vehicle ownership (if annual mileage exceeds approximately 15,000 km), and subsequent increases in real income. Based on the TCO calculation (see "Analysis of Incentives and Total Cost of Ownership in Bhutan" in chapter 5), it is estimated that reduction in fuel expenses will be about Nu 39,510 (US$659) per year for private vehicles (table 8.2).

Private taxis are another target group for EV replacement. Taxis carry over 36,000 passengers per day and are the principal mode of public transport in the city, accounting for about 30 percent of the mode share. The benefits for this group will be reduction in operating expenses, reduction in total cost of vehicle ownership, and subsequent increases in real income. For taxi drivers, fuel savings will be much higher at about Nu 213,300 (US$3,555) per year (table 8.2). The conversion of taxis to EVs will directly benefit existing taxi users financially if it results in cheaper fares. However, cheaper fares are unlikely to be achieved in the short term. Moreover, targeting taxis is another opportunity to allow the majority of the urban population to access EV. Given the key role of taxis in serving

Table 8.2 Estimated Annual and Monthly Fuel Expenses for ICE and Electric Vehicles

	Annual expenses		Monthly expenses	
	Nu	US$	Nu	US$
Personal vehicles				
ICE fuel expenses (10,000 km/year)	44,100	735	3,675	61
EV energy expenses (10,000 km/year)	4,590	77	380	6
Savings	39,510	659	3,295	55
Taxis		0		0
ICE fuel expenses (50,000 km/year)	236,250	3,938	19,670	328
EV energy expenses (50,000 km/year)	22,950	383	1,910	32
Savings	213,300	3,555	17,760	296

Source: World Bank analysis.
Note: Assumptions for the estimates are provided in appendix C. EV = electric vehicle; ICE = internal combustion engine; Nu = Bhutanese ngultrum.

tourist trips, targeting taxis has a potential indirect benefit of promoting the country's image as a green and environmentally friendly place.

The third target for EV replacement is the government fleet. The government currently owns 2,524 light vehicles. Replacement of this fleet with EVs will directly contribute to lower government expenses on fuel and a reduction in the budget allocation for vehicle operating expenses. Indirectly, replacing the government fleet will also have a symbolic benefit by demonstrating the government's commitment to the EV program. However, if the mileage for a vehicle is not high enough, the total cost of ownership (TCO) may be higher for an EV than a normal ICE vehicle. Table 8.3 summarizes the potential beneficiary groups, their economic characteristics, and their benefits. Box 8.1 provides more details on potential EV buyers in Bhutan.

In addition to obtaining individual benefits, EV buyers also generate positive externalities for the public, including improved urban air pollution, avoided GHG emissions, and reduced fuel imports. These benefits are enjoyed by the entire Bhutanese population.

Distributive Analysis

Several aspects of the EV program can have a distributive effect. This involves not only the buyers of the EVs, but also the replacement of the government fleet and the way fiscal incentives are used.

Table 8.3 Summary of Potential Beneficiaries and Benefits

Target group	Number of beneficiaries	Characteristics	Benefits
Private vehicles	• Number of eligible vehicles that may be suitable for EV replacement is around 6,750.[a] • Target for replacement in high EV uptake scenario is 969 EVs by 2020.	• High-income and upper-middle income group	• Lower monthly fuel expenses • If annual mileage is high enough, total cost of ownership (TCO) for the vehicle is reduced
Taxis	• Over 2,000 taxis in Thimphu • 33,000 passengers per day • Target for replacement in high EV uptake scenario is 441 EVs by 2020.	• Taxi drivers and users are from middle and lower-middle income group	• Lower monthly fuel expenses • If annual mileage is high enough, TCO is reduced • Net benefits will depend on impact on revenue and operating cost • Access to EV for majority of urban population
Government vehicles	• Around 2,500 government-owned light vehicles • Target for replacement in high EV uptake scenario is 62 EVs by 2020.		• Reduced budget on fuel expenses • Budget could be used for more productive purposes

Source: World Bank analysis.
Note: EV = electric vehicle.
a. Among the light vehicles for private use, half are SUVs or vehicles with engine displacement > 2,000 cc; for these vehicles currently no suitable EV offering is available.

Box 8.1 Who Are the EV Buyers in Bhutan?

Since EVs were introduced in the Bhutan market in July 2014, a number of vehicles have been sold. By mid-August 2014, the number of purchases was 27 vehicles, of which 2 were second-hand cars that are eligible only for taxi use. The majority of the buyers were businesspeople (50 percent), civil servants (11 percent), professionals (7 percent), housewives (4 percent), and corporations (20 percent). Among these groups, it appears that the main groups of EV buyers are from the upper-income (for example, business owners) and middle-income households (professionals and civil servants).

Why do customers buy EVs in Bhutan?
Based on short interviews with a few EV buyers, potential buyers, and EV sales offices, the main reasons for the early adopters to buy an EV include:

- Personal preference (curiosity, excitement for new product)
- Financial (getting a good deal from the subsidized price of an EV)
- Suitability for household car use patterns (short distance, within-city use, family second car)
- Environmental benefits and zero tailpipe emission
- Availability of parking space and home charging
- Substantial reduction in fuel expenses

In contrast, the key reasons for potential buyers not to make the purchase include:

- Lack of knowledge and reliable information on EVs
- Concerns about battery durability, battery second life, and availability of spare parts
- Unavailability of parking (for people living in public houses and residences with no designated parking) and charging (particularly among taxi drivers who need fast charging during the day)
- Concerns on actual life cycle CO_2 impact compared to ICE vehicles
- High upfront cost
- Needing a loan from a bank
- Waiting for feedback from other customers

According to the characteristics of vehicle owners, EVs are likely to be bought by higher-income households. The car is already a major mode of transportation among upper-middle- and higher-income households in the urban area. Over 50 percent of urban households in the fourth and fifth consumption quintiles own a family car compared to 35 percent of urban households in the third quintile (table 8.4). EV buyers will come from these consumption quintiles that are more likely to purchase a family car. In contrast, the distributive impact from the EV initiative is likely to be more positive in the taxi segment, as taxi drivers and users generally come from the lower-middle-income group.

Because of the high upfront cost of EVs, it is likely that EVs will be affordable only to top-income households. This means that, in addition to target segments,

Table 8.4 Distribution of Households That Own Transport Assets by Per Capita Consumption Quintile (Percentage)

Type of assets	Per capita consumption quintile					
	First	Second	Third	Fourth	Fifth	Total
Family car	12.7	22.2	34.5	50.3	57.7	**35.5**
Bicycle	6.1	8.2	7.6	12.1	12.8	**9.4**
Other vehicle	3.2	5.1	8.3	9.2	13.2	**7.8**
Motorbike, scooter	2.9	4.6	5.7	6.5	5.3	**5**

Source: National Statistics Bureau, http://www.nsb.gov.bt/publication/files/pub1tm2120wp.pdf.

Table 8.5 Annual Household Income Distribution

	Quintile					
	First	Second	Third	Fourth	Fifth	Average
Annual urban household income by quintile (Nu)	90,000	124,000	150,000	186,520	324,000	282,671
Annual urban household income by quintile (US$)	1,497	2,062	2,495	3,102	5,389	4,701

Source: National Statistics Bureau, http://www.nsb.gov.bt/publication/files/pub1tm2120wp.pdf.

Table 8.6 Estimates of Household Affordability for EV Purchase

	Taxis		Personal vehicles	
	ICE	EV	ICE	EV
Vehicle upfront cost after taxes Nu (US$)	558,000 (9,300)	800,000 (13,333)	806,000 (13,433)	1,260,000 (21,000)
Estimated monthly loan repayment Nu (US$)	6,700 (112)	9,600 (160)	9,800 (163)	15,292 (255)
4th quintile monthly income Nu (US$)	15,550 (259)	15,550 (259)	15,550 (259)	15,550 (259)
Loan repayment as % of monthly income	43%	61%	63%	98%
5th quintile monthly income Nu (US$)	27,000 (450)	27,000 (450)	27,000 (450)	27,000 (450)
Loan repayment as % of monthly income	25%	35%	36%	57%

Source: World Bank analysis.
Note: EV = electric vehicle; ICE = internal combustion engine; Nu = Bhutanese ngultrum.

affordability also is a key factor in determining who will be the beneficiaries of the EV program. Loan repayment ability in comparison with income can provide an indication about the level of affordability for a potential consumer. Because of the limited information on the income of potential EV buyers, only a preliminary analysis was made using information on household income for different quintiles from the Bhutan Living Standards Survey in 2012 (table 8.5). On the basis of this information, table 8.6 then shows a rough calculation of loan repayments (assuming a loan amount that is 50 percent of vehicle price, has a 14 percent interest rate, and has a five-year term).

The analysis shows that affordability is clearly an issue for the purchase of EVs. For example, loan repayment for the EV will take 61 percent of taxi drivers' average monthly income if their income is about Nu 15,500 (US$259) per month and 35 percent with income of about Nu 27,000 (US$450) per month. For personal vehicles, the loan repayment in the case of EVs will be 98 percent of monthly income for households in the fourth quintile and 57 percent in the fifth quintile. Because of the high upfront cost of an EV, it is likely that only the urban population in the fifth quintile will be able to afford an EV. Further data on the socioeconomic characteristics of potential target EV buyers are needed to better assess EV affordability.

In addition to the fact that EVs are affordable only to the higher income quintiles, the benefits of fuel savings will also disproportionally benefit the upper quintiles compared to the lower ones. The share of monthly transport-related expenses accounts for a sizable share (almost 20 percent for the urban area in general) of monthly household nonfood consumption expenditure (table 8.7 and figure 8.1), and this share is higher in the upper consumption quintiles than in the lower ones. In the fifth consumption quintile, the share of monthly transport and communication expenses was the highest at 26 percent of mean per capita nonfood household consumption expenditure (figure 8.1).

For the government fleet, the distributive implication of a change to EV will depend on whether the replacement of the fleet will result in a net savings of budget allocated for government transportation expenses (that is, if a lower TCO is achieved) and on how the savings will be used, for example, for which public services program.

Finally, the use of fiscal incentives to promote EVs can also have strong distributive implications. The increases in green tax, sales tax, and customs duty tax will reduce the number of car purchases, particularly among lower-income groups because higher taxes reduce their ability to afford a car. The higher-income groups that can afford EVs can enjoy the benefit of a 90 percent tax exemption

Table 8.7 Mean Monthly Per Capita Household Food and Nonfood Expenditure by Per Capita Household Consumption Quintile in the Urban Area (Nu)

Per capita household consumption expenditure quintile	Mean per capita expenditure	Food	Health	Education	Miscellaneous	Transport and communications	Rent	Other nonfood
Urban	5,804.1	1,964.9	273.2	185.9	349.7	740.1	881.7	1,409.0
First	2,480.7	1,191.5	41.4	83.0	135.1	147.7	342.1	540.0
Second	3,802.4	1,625.0	86.0	121.7	230.3	292.7	583.2	863.6
Third	5,089.2	1,939.9	127.9	204.8	330.3	495.0	810.7	1,180.7
Fourth	7,087.1	2,408.3	298.1	295.3	448.3	784.3	1,146.1	1,706.6
Fifth	14,139.5	3,312.6	1,151.9	296.7	812.9	2,778.7	2,054.4	3,732.3

Figure 8.1 Share of Mean Monthly Per Capita Household Transport and Communication Expenses of Total Per Capita Household Nonfood Consumption by Quintile in Urban Area

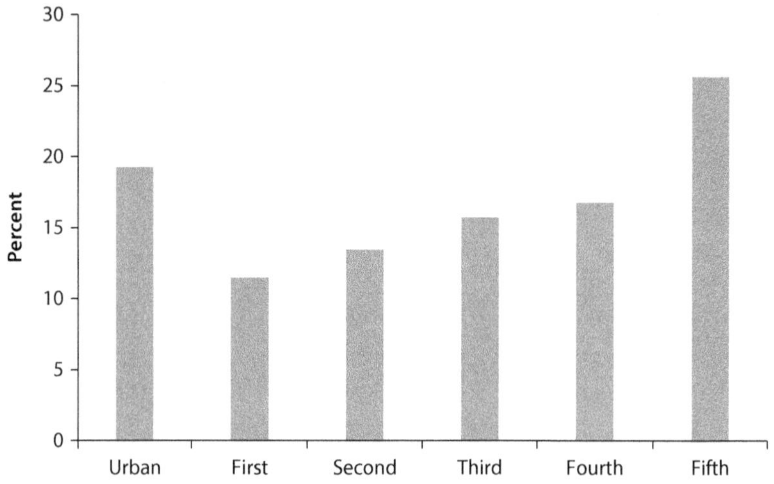

Source: Asian Development Bank and National Statistics Bureau of Bhutan 2013.

on those vehicles. The increased tax rates, however, will improve tax collection and increase government revenue, with increased vehicle taxes generally collected from middle- and higher-income groups. Use of this additional tax revenue will also have distributive implications. In general, if these taxes are used for public transport, it can benefit bus users like students and lower-income groups. If the EV program requires additional government subsidy, for example for the operation of charging infrastructure, the taxes collected from those who purchased an ICE vehicle will benefit EV buyers who are already exempt from taxes.[1] Any additional subsidy under the EV program should be carefully reviewed to avoid cross subsidy from other groups of taxpayers, while also considering the government's fiscal space and competing priorities for public funds.

Overall, the distributive impact of the EV program and broader public benefits resulting from it need to be weighed carefully. The distributive analysis suggests that the EV initiative is likely to benefit higher-income households more than lower-income households, and distributive implications are likely to be negative. However, as will be discussed in "EV in a Broader Context: The Role of Public Transport in Green Mobility," below, the EV initiative brings public benefits in terms of reduced fuel consumption and environmental benefits (see also chapter 7). The cost of providing fiscal incentives to EV buyers should then be assessed against the positive externalities generated by EV consumers to ensure that the green benefits and equitable aspect of the policy are in balance. Table 8.8 shows basic estimates of public benefits and public costs for the three scenarios. To carry out a full cost-benefit analysis, additional information will need to be collected.

Table 8.8 Estimates of Public Benefits and Costs

EV uptake scenario	Low EV uptake, in million Nu (US$)	High EV uptake, in million Nu (US$)	Super high EV uptake, in million Nu (US$)
Public benefits			
Accumulated avoided fuel imports (2015–2027)	768 (13)	2,479 (41)	18,531 (309)
Accumulated avoided GHG emissions (2015–2027)	31 (1)	100 (2)	750 (13)
Total public benefits	799 (13)	2,580 (43)	19,282 (321)
Public costs			
Total fiscal cost of incentives program (2015–2020)	593 (10)	1,967 (33)	7,140 (119)
Total investment in public infrastructure (2015–2020)	62 (1)	320 (5)	752 (13)
Total procurement of government fleet replacement (2015–2020)	32 (1)	96 (2)	160 (3)
Total public costs	687 (11)	2,383 (40)	8,052 (134)

Source: World Bank analysis.
Note: Assumptions for the estimates are provided in appendix C. EV = electric vehicle; GHG = greenhouse gas; Nu = Bhutanese ngultrum.

EV in a Broader Context: The Role of Public Transport in Green Mobility

Bhutan's projected economic growth will accelerate the acquisition of cars by the country's growing middle class, and the introduction of EVs presents an opportunity to reduce the country's dependence on energy imports, along with a reduction in the climate and air pollution impacts of the growing motorization. Most other countries with EV initiatives introduce EV as part of a broader green transport effort. In Bhutan, improving public transport in parallel with an EV policy will allow the country to achieve green mobility in a way that addresses the mobility needs of the entire population.

Public Transport in Bhutan

At the national level, long distance and interprovincial bus services are provided by the private sector; these services rely on self-financing. Passengers on these buses generally do not have access to private transport and typically are from lower-income households. An exception within this bus system is the Thimphu Bus Services. The services are operated by Bhutan Postal Corporation for the Ministry of Information, Communications, and Technology and receive an operating subsidy from the Ministry of Finance. The subsidy per bus operated has been gradually declining over the past few years, and the current main concern for the service is obtaining more bus capacity to meet the latent and rising demand for public transport services.

In addition to the official buses, in practice many taxis operate as small and unscheduled bus services, charging separate fares per passenger rather than being hired by a person or group. For longer distance interprovincial travel, buses are much cheaper than taxis and therefore more popular, although taxis (often as collective bus services) are also used. In urban areas such as Thimphu, the absolute difference in cost between bus and taxi travel is relatively low, especially when passengers can share taxi rides. As a result, taxis carry about five times the number of passengers carried by buses within Thimphu, at least in part because of insufficient bus system capacity relative to demand (see also appendix A).

A related trend that likely reflects the growth in motorization since 2003 is the decline in walking. In 2003 walking still represented 60 percent of all trips made throughout the nation, but this share declined to 20 percent in 2012 (Asian Development Bank and National Statistics Bureau of Bhutan 2013). While much of this reduction is likely a direct result of growing motorization, walking connections to bus stops and community services are often indirect and unprotected from traffic.

Improving Public Transport to Achieve Inclusive Green Transport

Improving public transport, along with the implementation of an EV initiative, presents an option to provide green transport in a way that caters to all levels of Bhutan's population. As the EV program has a distributive effect, care is needed to ensure government financial support for EVs is not to the disadvantage of low-income households and public transport users. The majority of the existing bus users, many of whom are schoolchildren, come from families who cannot afford to purchase ICE vehicles, let alone the more expensive EVs, or even to use a taxi on a regular basis. The mobility needs of this group very much depend on the availability of a reliable, basic bus service, with affordable fares and adequate coverage and frequency.

In the case of the Thimphu Bus Services, options exist to improve the bus services; some areas for improvement are listed in box 8.2. In addition, the bus technology—whether using diesel, hybrid, and electric bus technologies—can contribute to CO_2 emissions reduction. The cost of these technologies and their contribution to emissions reduction are analyzed in appendix F. When choosing a technology, the road situation, loading capacity, driving distances, and other bus operation characteristics will also need to be considered. Currently, electric buses are not considered to be well suited for Thimphu. Electric buses carry a substantial cost and are still an unproven technology for small urban environments such as Thimphu, with hilly terrain and limited technical capacity. In the short term, continued purchase of low-emission diesel buses can effectively address the current major challenge of providing new bus services to meet demand.

In addition to providing for a population that has no choice but to use public transport, improved services that efficiently connect homes to work and services may also attract "choice" users by providing substantial welfare benefits (for example, savings in time and out-of-pocket expenses) along with significant green benefits. Addressing public transport in a more systematic manner will also help address fuel dependency in the long run by inducing a modal switch from private cars even in the middle-income groups that are the main users of public

Box 8.2 Areas for Improvement of Thimphu Bus Services

An analysis of Thimphu Bus Services identified several areas for improvement.

Short term (next two years):
- Staff training
- Improved spare parts supply chain
- Improved service monitoring and reporting
- Provision of better bus stops and shelters
- Improved marketing of services
- Increased bus fleet (conventional diesel buses)
- Improved service frequencies
- Changing Norzin Lam into a two-way bus-only street
- Review of current land use development plans to ensure provision of infrastructure for efficient bus services and pedestrian movement
- Preparation of Bus System Master Plan

Longer Term:
- Adoption of smart card ticketing
- Real time bus monitoring based on Global Positioning System (GPS)
- New services to new development areas
- Further bus priority traffic management measures
- Larger/improved maintenance depot facilities
- City bus services to be transferred to Thimphu Thromde
- Physical improvements to the city bus terminal

transport systems, as seen in other cities in developing countries. If—in the long run—buses are converted to EVs, this could further improve the quality of the ride and possibly (depending on vehicle capital costs and alternative fuel/power costs) reduce fares. Although a switch from taxi use to bus use could be considered a negative impact on taxi operators, given the anticipated ongoing rapid rise in population and overall travel demand, this is more likely to slow the rate of growth in taxi use rather than represent an absolute reduction.

Note

1. Because the new tax regime was only recently put in place, the assessment is only an initial one. It is important to see how the market responds to the new tax regime to better assess the distributive implications of the fiscal incentive.

Reference

Asian Development Bank, and National Statistics Bureau of Bhutan. 2013. *Bhutan Living Standards Survey 2012 Report*. Mandaluyong City, Philippines: Asian Development Bank.

APPENDIX A

Background Information on Urban Transport in Bhutan

The following is a very brief overview of urban transport in Bhutan.

Bus Services and Taxi Use

In FY2011/12, the inter-Dzongkhag passenger service (Thimphu Bus Service) carried over 1,010,924 customers, which was 13.08 percent more than the previous year. However, less than a third of urban households in Bhutan reported using public transport in a given month. In Thimphu, for example, city buses carry approximately 6,000 passengers per day, while taxis carry approximately 36,000 passengers per day. Many of the taxi passengers during peak periods are being carried on a "shared ride, fixed individual fare" basis. A clear scope exists to reduce peak-period taxi traffic in Thimphu by diverting taxi trips to bus travel, with consequent improvements to traffic conditions.

Private Cars

Bhutan's urban population is increasing (figure A.1) and along with it motor vehicle ownership and fuel imports (figure A.2).

As of September 2014, the total number of vehicles registered in Bhutan was 68,744. Based on household ownership rates, it is estimated that about 38.4 percent of vehicle registration is in Thimphu Dzongkhag. The motorization rate grew about 12 percent per year during 2008–2012 (see figure A.3) with an average of about 6,300 newly registered vehicles each year. Light vehicles and taxis, both privately and publicly owned, account for 70 percent of the vehicle mix (see table A.1). As shown in the table, it is estimated that there are about 15,075 light vehicles in Thimphu Dzongkhag, with an additional 2,500 vehicles registered each year. In 2013, the country had a total of 5,145 taxis, of which 2,024 were in Thimphu.

Figure A.1 Bhutan Urban Population

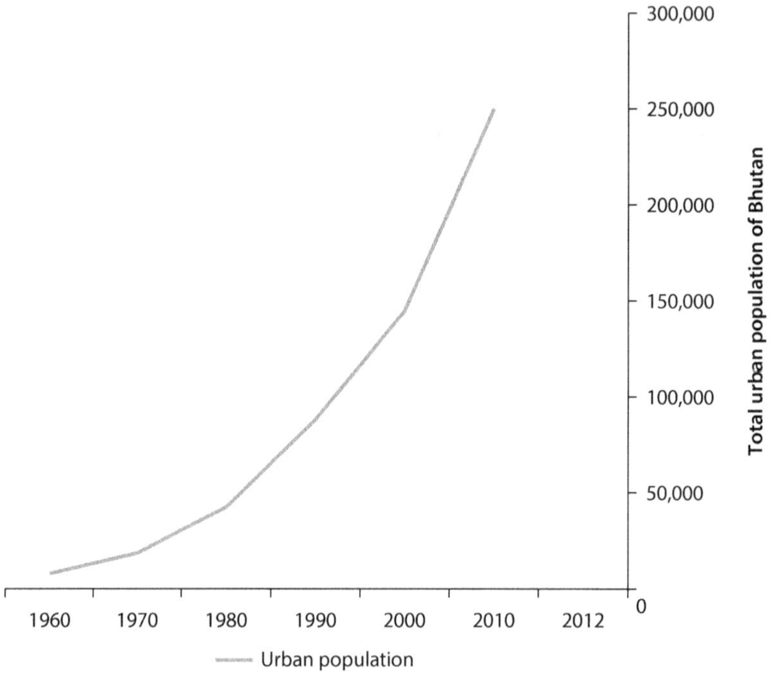

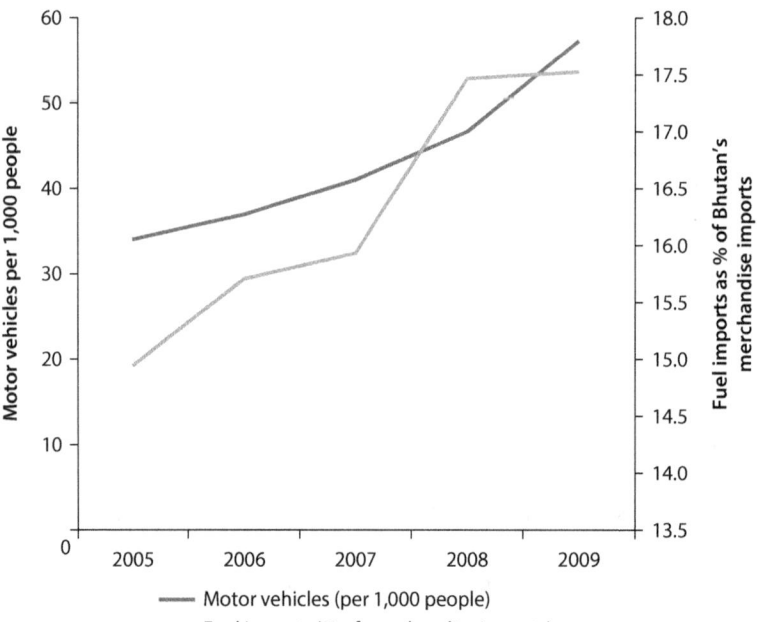

Source: World Bank Development Indicators.

Figure A.2 Motor Vehicle Ownership in Bhutan

Source: World Bank Development Indicators.

Background Information on Urban Transport in Bhutan 139

Figure A.3 Vehicle Registration in Bhutan, 1997–2013

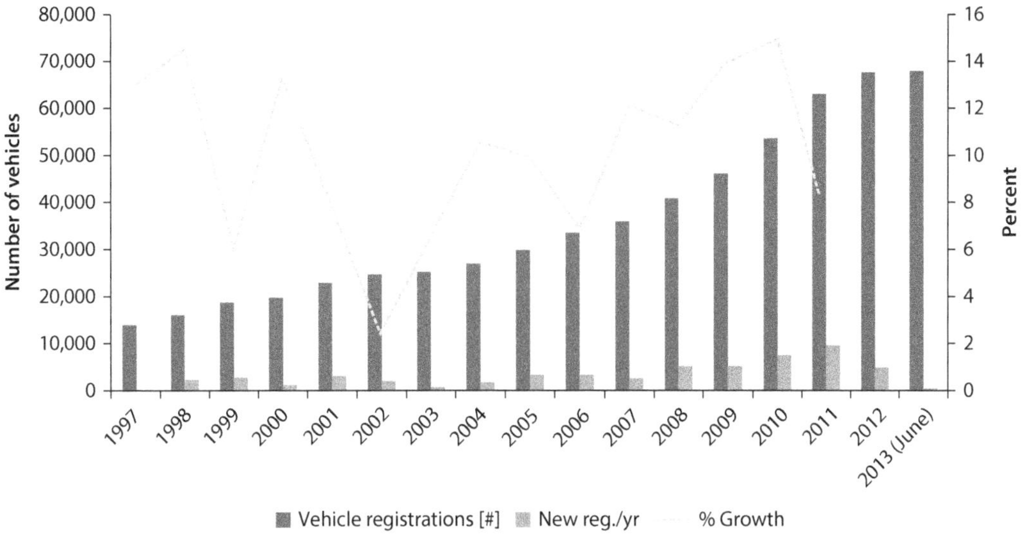

Source: http://www.rsta.gov.bt/download/AnnualReport2012-2013.pdf.

Table A.1 Composition of the Vehicle Fleet in Bhutan

Vehicle class	Government	Private	Total	Thimphu region (estimate)	Thimphu Dzongkhag (estimate)
Light	2,524	36,736	39,260	20,889	15,075
Two wheeler	1,466	8,319	9,785	5,206	3,757
Heavy	1,164	7,383	8,547	4,547	3,282
Taxi	0	5,271	5,271	2,804	2,024
Equipment	250	1,703	1,953	1,039	750
Medium	232	1,112	1,344	715	516
Power tiller	58	1,151	1,209	643	464
Tractor	89	229	318	169	122
BHT (LV)	0	69	69	37	26
CD (LV)	0	63	63	34	24
CD (TW)	0	5	5	3	2
Total	5,783	62,041	67,824	36,086	26,042

Source: http://www.rsta.gov.bt/download/AnnualReport2012-2013.pdf.
Note: BHT (LV) = Bhutan light vehicle (for the royal family); CD (LV) = country diplomat light vehicle; CD (TW) = country diplomat two-wheeler.

APPENDIX B

Examples of International Incentive Programs

Several countries have implemented incentive programs to promote the uptake of electric vehicles (EVs). Table B.1 presents an overview of the specific incentives—tax credits, charges, exemptions, and subsidies—in selected countries. Additional details for each country are provided below the table.

China

- In 2009, China initiated a Ten Cities, Thousand Vehicles Program. The program uses large-scale pilots to address technology and safety issues. Each city was challenged to roll out at least 1,000 vehicles, with a focus on government fleet vehicles with predictable driving patterns such as buses, garbage trucks, and taxis. In 2010, the program expanded to include consumers.
- As listed in table B.1, the central government introduced purchase subsidies of Y 60,000 per vehicle for battery electric vehicles (BEVs) and Y 50,000 for PHEVs. In addition, state-level subsides are available.
- In Shenzhen, additional subsidies of Y 60,000 for BEVs and 20,000 for PHEVs are being offered. In this case, the total consumer purchase subsidies are Y 120,000 and Y 70,000 for BEVs and PHEVs, respectively.
- Central and local government investments amount to Y 100 billion.

Japan

- Key incentive for consumer is a price gap between internal combustion engine (ICE) vehicles and EVs of up to ¥1 million (US$10,000) per vehicle and half the cost of public chargers up to ¥1.5 million (US$15,000) per charger.
- As an industrial leader, government has provided major support to research and development (R&D).

Table B.1 EV Incentive Programs in China, India, Japan, the United Kingdom, and the United States (2014)

Country	Tax credits	Tax and charges exemption	Subsidies
China			• Y 60,000 (Chinese Yuan) per vehicle for BEVs and Y 50,000 for PHEVs. • Additional subsidies at the state level.
India		Reduced excise duties on BEV/PHEVs.	• Rs 100,000 or 20% of cost of vehicle whichever is less.
Japan			• Support to pay for half of the price gap between EV and corresponding ICE vehicles, up to ¥1 million per vehicle.
United Kingdom		• Private electric vehicles exempt from annual circulation tax. • Company electric cars exempt from company car tax for first five years after purchase. • Exemption from congestion charging.	• Subsidy of 25% of vehicle list price with a cap of £5,000. • Subsidy on home charging station.
United States	• Up to US$7,500 consumer tax credits for new purchase of PHEV/EV based on battery capacity. Phased out after 200,000 vehicles from qualified manufacturers. • Additional state-level purchase incentives up to US$5,000 for PHEV/EV.		

Note: BEV = battery electric vehicle; EV = electric vehicle; ICE = internal combustion engine; PHEV = plug-in hybrid electric vehicle.

Norway

- Norway has the highest per capita rate of PHEV ownership in the world.
- Climate policy is a major driver. A greenhouse gas (GHG) and other emissions component has been applied to the first-time annual registration charge, which has resulted in a doubling of prices of new ICE cars.
- Other incentives include a waiver of 25 percent value added tax (VAT), exemption from registration taxes and tolls, free parking, use of bus lanes, and free downtown charging.
- The current EV incentive policy is the result of more than 20 years of experience in adapting the measures to the response of the population.

The United Kingdom

- £350 million has been made available for research and demonstration projects.
- A £20 million procurement program has been planned for government vehicles; 25,000 charging points will be installed in London.

- As listed in table B.1, private vehicles are exempted from annual circulation tax; company cars are exempted from the tax for company cars.
- Consumer purchase subsidy of 25 percent of vehicle list price with a cap of £5,000; the government has set aside £230 million for the incentive.
- Subsidy on approved home charging stations.

The United States

- To stimulate manufacturing and R&D investment, the government has invested US$25 billion for an Advanced Technology Vehicle Manufacturing Incentive program to develop technology that achieves 25 percent higher fuel economy and US$2.4 billion grants for EV development.
- For infrastructure investment, tax credits of US$54 million on alternative refueling property including charging are provided, as well as a US$100 million grant for five-city infrastructure deployment.
- Consumer tax credits and purchase incentives are provided for PHEV (plug-in hybrid electric vehicle)/EV purchase. In addition, consumers also enjoy high-occupancy vehicle (HOV) lane access; designated parking space programs are provided by the states.

APPENDIX C

Total Cost of Ownership Analysis for Bhutan

Introduction to the Total Cost of Ownership Analysis

Generally, the aim of fiscal incentives is to reduce the total cost of ownership (TCO) of an electric vehicle (EV) in comparison with the TCO of an internal combustion engine (ICE) vehicle. This study used the approach of a TCO cost comparison between electric and ICE vehicles to estimate the potential for switching, as is done in other countries. Analyses of TCO and upfront costs are useful because consumers often make purchase decisions by considering both capital and operating costs. Not all customers, however, may value upfront and operating costs in the same way (or even rationally).

The fiscal incentives included in this study are limited to the use of taxes and cost subsidies, as requested by the government. Fiscal incentives are calculated for different EV scenarios based on the assumptions on how much incentive will be needed to meet the EV targets for each uptake scenario.

Assumptions for the TCO Analysis

Assumptions, including the selection of vehicles to compare, taxes, and assumed fuel price increases, are listed in the tables below for private vehicles, taxis, and government vehicles (tables C.1, C.2, C.3).

Results of TCO Analysis for Private Vehicles

Using the assumptions described above, table C.4 lists the TCO for EVs and ICE vehicles based on the analysis, for three discount rates.

Scenario Analysis for Setting Incentives—Private Vehicles

Similar to the situation in other countries, it is assumed in Bhutan that the early EV adopter groups are not be influenced by TCO; but as the market grows the majority of buyers will be influenced increasingly by the price factor. For this

Table C.1 Assumptions for the TCO Analysis for Private Vehicles

Variables	Values	Remarks
EV fleet composition	Nissan Leaf comprises 100% of total EV fleet	For the purpose of this analysis, one EV model is compared with a comparable ICE model. Given the local market development, the Nissan Leaf (high-end model), which is available in the local EV market at a 30% subsidized price of Nu 1,260,000 by the manufacturer, is selected as a representative EV. Information on prices for the marginal vehicle after the first lot is not available.
Comparable ICE	Hyundai i20 at Nu 520,000	Hyundai i20 is selected as a comparable ICE vehicle. Its product dimensions are relatively close to those of the Nissan Leaf and the car is available in the local market.
EV model	Nissan Leaf at Nu 1,260,000	
Vehicles taxes on ICE	Sales tax 45% Green tax 10%	The new effective tax rates for vehicles below 1,500 cc as of July 2014 are show in table 5.2.
Vehicle taxes on EV	0%	The exemption of sales tax, green tax and customs duty is the current fiscal incentive to promote EVs.
Average distance travelled per year	10,000 km	Estimate for Bhutan.
Fuel economy—ICE	15 km/liter	Estimate for Bhutan.
Fuel economy—EV	5.36 km/kWh	U.S. Environmental Protection Agency.
Fuel price	Nu 63	As of July 2014.
Fuel tax	5%	In July 2014, the government also started to impose 5% tax on fuel in an attempt to curb fuel growth and align fuel prices with India's.
Fuel price escalation p.a.	7%	GNHC assumption in a study "Cost Benefit Analysis of Introducing Electric Vehicle in Bhutan" based on Agarwal, Pradeep. 2012. "India's Petroleum Demand: Empirical Estimations and Projections for the Future." *IEG Working Paper*, No. 319.
		This rate is in line with (a) historical increases of Electricity, Gas, and Other Fuels Index in Bhutan based on Consumer Price Index (CPI) report at 8% p.a. during 2012–2014 and (b) Housing, Water, Electricity and Gas/Fuels Index in the CPI at 8.5% p.a. during 2004–2013. Data on historical fuel prices are not available. Note that the fuel price increase of 1% is tested at the lower bound. The assumption of 1% nominal fuel price increase is based on World Bank projection of crude oil average spot price (Brent, Dubai, West Texas Intermediate) which projected crude oil price at US$103.4 per barrel in 2025 (2014 price is US$96.2 per barrel) representing average annual increase of 1%. Other projections in 2012 dollars per barrel range from −3.3% to 3.2% changes p.a. during 2012–2025.
Electricity price	Nu 2.46	As of 2014 Block III tariff.
Electricity price escalation	14%	According to BPC tariff schedule, based on GNHC assumption.
Maintenance expenses per vehicle/year—cars	0.05 US$/km	Estimate
Maintenance expenses per vehicle/year—EVs	0.035 US$/km	Estimate
Years of usage	8 years	Estimate
Disposal value—ICE	10% of capital	Estimate
Disposal value—EV	10% of capital	Estimate

table continues next page

Table C.1 Assumptions for the TCO Analysis for Private Vehicles *(continued)*

Variables	Values	Remarks
Loan financing	50% of vehicle price	It is assumed that an average consumer will finance a vehicle through loan (50%) and his/her own saving (50%) based on the financing terms and conditions set out below. The ban on vehicle loan was lifted on September 1, 2014. Conditions: (a) 5-year loan, (b) 14% interest rate on vehicle loan, (c) 7% return on long-term deposit which represents opportunity cost of a consumer, and (d) WACC (Weighted Average Cost of Capital) for a vehicle purchase is (14%*0.5) + (7%*0.5) = 10.5%, rounded to 10%.
Loan term	5 years	Interview
Interest rate	14%	As of 2014
Exchange rate	60 Nu/US$	As of July 2014
Inflation p.a.	8%	Based on average 10 years historical CPI (2004–2013)
Discount rate	10%	The 10% discount rate reflects the cost of capital of a vehicle purchase for an average consumer (see above). 5 and 15% discount rates are also used for sensitivity analysis to provide lower bound and upper bound. Market reference for discount rate for investment projects in the local market is 10% based on hydropower project and 7% in general.
Emission	165 g/km	GNHC assumption
Social value of carbon	US$30 per ton of CO_2	Based on World Bank (2014) Social Value of Carbon Project Appraisal. US$30 per ton of CO_2 is the social value of carbon in the base case in real 2014 U.S. dollars.

Source: World Bank analysis.
Note: BPA = Bhutan Power Corporation; EV = electric vehicle; GNHC = Gross National Happiness Commission; ICE = internal combustion engine; kWh = kilowatt-hour; p.a. = per annum; TCO = total cost of ownership.

Table C.2 Assumptions for the TCO Analysis for Taxis

Variables	Values	Remarks
EV fleet composition	Nissan Leaf comprises 100% of total EV fleet	Among the locally available models (brand new Nissan Leaf, refurbished Nissan Leaf and Mahindra Reva e2o), the most suitable EV model for taxis is the refurbished Nissan Leaf (driven for about 20,000–30,000 km), for which the price (Nu 800,000) is affordable for the taxi drivers and which also has suitable capacity for taxi operations. Given that these vehicles may be made available in limited quantities, and given environmental concern of importing a secondhand vehicle, the calculation was made for both the refurbished Nissan Leaf and regular Nissan Leaf price.
Comparable ICE	Maruti Suzuki Alto at Nu 520,000	The Maruti Alto, which currently dominates the taxi fleet, is selected as the comparable ICE vehicle for taxis.
EV	Nissan Leaf Refurbished at Nu 800,000	
Vehicles taxes on ICE	Sales tax 45% Green tax 10%	The new effective tax rates for vehicles below 1,500 cc as of July 2014 are show in table 5.2.
Vehicle taxes on EV	0%	The exemption of sales tax, green tax, and customs duty is the current fiscal incentive to promote EVs.
Average distance travelled per year	50,000 km	Estimate

table continues next page

Table C.2 Assumptions for the TCO Analysis for Taxis *(continued)*

Variables	Values	Remarks
Battery replacement	Nu 165,000 after 150,000 km	See chapter 4.
Fuel economy—ICE	15 km/liter	Estimate
Fuel economy—EV	5.55 km/kWh	U.S. Environmental Protection Agency adjusted for refurbished Nissan Leaf
Fuel price	Nu 63	As of July 2014
Fuel tax	5%	In July 2014, the government also started to impose a 5% tax on fuel in an attempt to curb fuel growth and align fuel prices with India's.
Fuel price escalation p.a.	7%	See remarks for fuel price escalation in table C.1.
Electricity price	Nu 2.46	As of 2014 Block III tariff.
Electricity price escalation	14%	According to BPC tariff schedule, based on GNHC assumption.
Maintenance expenses per vehicle/year—cars	0.05 US$/km	Estimate
Maintenance expenses per vehicle/year—EVs	0.035 US$/km	Estimate
Years of usage	8 years	Estimate
Disposal value—ICE	10% of capital	Estimate
Disposal value—EV	10% of capital	Estimate
Loan financing	50% of vehicle price	See remarks for loan financing in table C.1.
Loan term	5 years	Interview
Interest rate	14%	As of 2014
Exchange rate	60 Nu/US$	As of July 2014
Inflation p.a.	8%	Based on avg. 10 years historical CPI (2004–2013)
Discount rate	10%	The 10% discount rate reflects the cost of capital of a vehicle purchase for an average consumer (see above). 5 and 15% discount rates are also used for sensitivity analysis to provide lower bound and upper bound. Market reference for discount rate for investment projects in the local market is 10% based on hydropower project and 7% in general.
Emission	165 g/km	GNHC's assumption
Social value of carbon	US$30 per ton of CO_2	Based on World Bank (2014) Social Value of Carbon Project Appraisal, 30 USD per ton of CO_2 is the social value of carbon in the base case in real 2014 U.S. dollars.

Source: World Bank analysis.
Note: BPA = Bhutan Power Corporation; EV = electric vehicle; GNHC = Gross National Happiness Commission; ICE = internal combustion engine; kWh = kilowatt-hour; TCO = total cost of ownership.

Table C.3 Assumptions for the TCO Analysis for Government Vehicles

Variables	Values	Remarks
EV fleet composition	Nissan Leaf comprises 100% of total EV fleet	For the purpose of this analysis, one model of EV is compared with a comparable model of ICE. Given the local market development, the Nissan Leaf (high-end model), which is available in the local EV market at a 30% subsidized price of Nu 1,260,000 by the manufacturer, is selected as a representative EV. Information on prices for the marginal vehicle after the first lot is not available.

table continues next page

Table C.3 Assumptions for the TCO Analysis for Government Vehicles *(continued)*

Variables	Values	Remarks
Comparable ICE	Hyundai i20 at Nu 520,000	Hyundai i20 is selected as a comparable ICE vehicle. Its product dimensions are relatively close to those of the Nissan Leaf and the car is available in the local market.
EV	Nissan Leaf at Nu 1,260,000	
Vehicle taxes on ICE	Sales tax 45% Green tax 10%	The new effective tax rates for vehicles below 1,500 cc as of July 2014 are shown in table 5.2.
Vehicle taxes on EV	0%	The exemption of sales tax, green tax and customs duty is the current fiscal incentive to promote EVs.
Average distance travelled per year	10,000 km	Estimate for Bhutan
Fuel economy—ICE	15 km/liter	Estimate for Bhutan
Fuel economy—EV	5.36 km/kWh	U.S. Environmental Protection Agency
Fuel price	Nu 63	As of 2014
Fuel tax	5%	In July 2014, the government also started to impose a 5% tax on fuel in an attempt to curb fuel growth and align fuel prices with India's.
Fuel price escalation p.a.	7%	See remarks for fuel price escalation in table C.1.
Electricity price	Nu 2.46	As of 2014 Block III tariff.
Electricity price escalation	14%	According to BPC tariff schedule, based on GNHC assumption.
Maintenance expenses per vehicle/year—cars	0.05 US$/km	Estimate
Maintenance expenses per vehicle/year—EVs	0.035 US$/km	Estimate
Years of usage	8 years	Estimate for Bhutan
Disposal value—ICE	10% of capital	Estimate
Disposal value—EV	10% of capital	Estimate
Loan financing	50% of vehicle price	No loan financing for government fleet; purchase is financed by national budget.
Loan term	5 years	Interview
Interest rate	14%	As of 2014
Exchange rate	60 Nu/US$	60 Nu/US$
Inflation p.a.	8%	Based on average 10 years historical CPI (2004–2013)
Discount rate	10%	The 10% discount rate reflects the cost of capital of a vehicle purchase for an average consumer (see above). 5 and 15% discount rates are also used for sensitivity analysis to provide lower bound and upper bound. Market reference for discount rate for investment projects in the local market is 10% based on hydropower project and 7% in general.
Emission	165 g/km	GNHC's assumption.
Social value of carbon	US$30 per ton of CO_2	Based on World Bank (2014) Social Value of Carbon Project Appraisal. 30 USD per ton of CO_2 is the social value of carbon in the base case in real 2014 U.S. dollars.

Source: World Bank analysis.
Note: BPA = Bhutan Power Corporation; EV = electric vehicle; GNHC = Gross National Happiness Commission; ICE = internal combustion engine; kWh = kilowatt-hour; TCO = total cost of ownership.

Table C.4 TCO and Switching Values—Private Vehicles

	Discount rate 10%	Discount rate 5%	Discount rate 15%
TCO ICE	(1,307,363)	(1,454,469)	(1,189,524)
30% subsidized Nissan Leaf			
TCO EV base case	1,436,676	1,514,136	1,364,158
Saving base case	(129,313)	(59,667)	(174,634)
Switching value—cost of ICE after tax	945,500	992,000	868,000
Switching value—annual mileage	15,000	12,000	17,000
Full price Nissan Leaf			
TCO EV base case	1,982,875	2,075,447	1,892,574
Saving base case	(675,512)	(620,978)	(703,049)
Switching value—cost of ICE after tax	1,457,000	1,395,000	1,519,000
Switching value—annual mileage	33,000	38,000	27,000

Source: World Bank analysis.
Note: EV = electric vehicle; ICE = internal combustion engine; kWh = kilowatt-hour; TCO = total cost of ownership.

group, the stronger the incentives are, the higher the EV uptake rate will be. Thus, for the fiscal analysis, hypothetical assumptions are made on the percentage difference between the TCO of ICE and electric vehicles that will induce switching from ICE to EV in the three scenarios (i.e., a higher target will demand stronger incentives to lower the TCO of an EV to further stimulate demand).

The hypothetical assumptions for each scenario are the following:

- Under the low EV uptake scenario, only highly price-inelastic consumer groups (for example, the top income bracket or early adopters) will buy an EV even if the TCO is much higher than the TCO for ICE vehicles. There is no need for additional incentives. This corresponds to a relatively low EV uptake target.
- Under the high EV uptake scenario, to meet the higher EV target, TCO of EVs should be less than or equal to the TCO of ICE vehicles to induce purchases among more price-elastic consumer groups compared to the first group. The additional incentive of cost subsidy is required to bring down TCO.
- Under the super high EV uptake scenario, to meet a more ambitious EV target, stronger financial incentives will be needed to further drive down cost. It is assumed that the target can be reached if the TCO of EVs is significantly below the TCO of ICE vehicles, such as by 20–30 percent, in order to attract a highly price-elastic consumer group.

For this analysis, calculations are made only for the base scenario using a 10 percent discount rate and the subsidized price of a Nissan Leaf. The result of the analysis suggests that with the new tax regime, the level of cost subsidy required to meet the target is 10 percent for the high scenario and 35 percent for the super high scenario. The TCO and upfront cost comparison are provided in figures C.1 and C.2 and table C.5.

Figure C.1 TCO Comparison

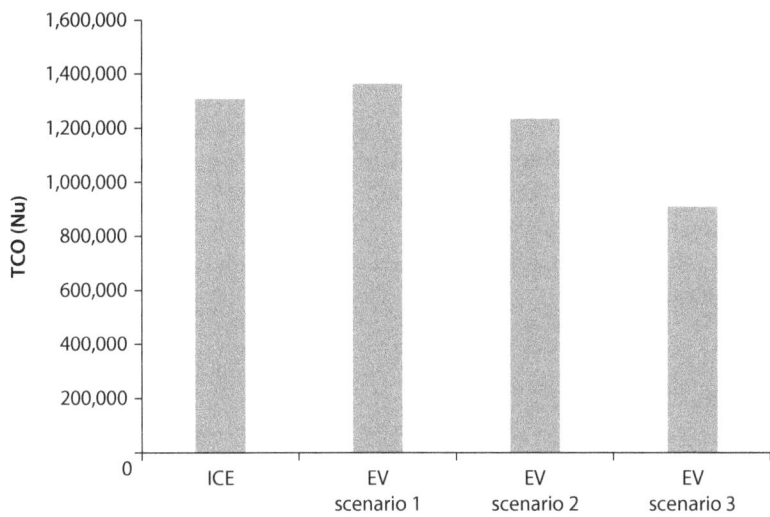

Source: World Bank analysis.
Note: EV = electric vehicle; ICE = internal combustion engine; TCO = total cost of ownership.

Figure C.2 Upfront Cost Comparison

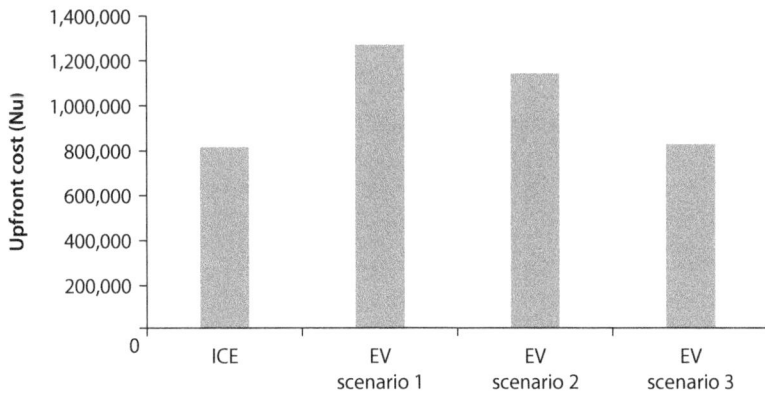

Source: World Bank analysis.
Note: EV = electric vehicle; ICE = internal combustion engine.

Table C.5 Cost Comparison of EV and ICE Vehicle (%) at 10% Discount Rate
Percent

	TCO	Upfront cost	Energy cost/month	Energy cost per km
EV and ICE scenario—low	4.3	56	−87	−90
EV and ICE scenario—high	−5.6	41	−87	−90
EV and ICE scenario—super high	−30.6	2	−87	−90

Source: World Bank analysis.
Note: EV = electric vehicle; ICE = internal combustion engine; TCO = total cost ownership.

Results of TCO Analysis for Taxis

The TCO analysis for the taxis segment is preliminary because it takes into account only the cost side of the taxi operation; additional information is needed to estimate any potential revenue impact of shifting to EVs. Revenue impact could result from lost time as a result of having to charge several times of day, representing a loss in revenue for each driver. In addition, a more detailed TCO analysis will also need to take into account different vehicle use patterns, which will be unique in the taxi segment, and other taxi-specific factors.

Because several factors related to the taxis are unknown, calculations considered several options:

- The cost of the EV at refurbished price (Nu 800,000) and full price (Nu 1,800,000)
- The option with the cost of battery replacement (annual mileage of 50,000 km, which will require a battery replacement costing Nu 165,000 in year three) and the option without battery replacement (annual mileage of 20,000 km, which will require no battery replacement in the 8-year operating period)

Figures C.3 and C.4 illustrate the TCO comparison (EV vs. ICE) for taxis for two different estimated annual fuel price increases (7 percent and 1 percent).

Based on the initial analysis, switching to EVs in the case of taxis will be financially viable. However, other factors will also influence this switch, given the

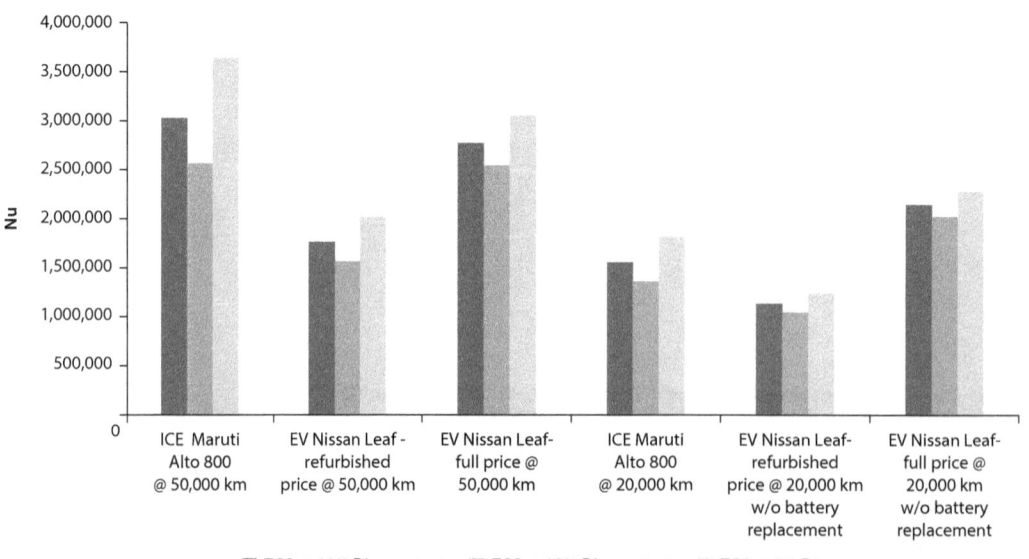

Figure C.3 TCO Comparison of ICE and EV for Taxis, Annual Fuel Price Increase at 7%

Source: World Bank analysis.
Note: EV = electric vehicle; ICE = internal combustion engine; TCO = total cost of ownership.

Figure C.4 TCO Comparison of ICE and EV for Taxis, Annual Fuel Price Increase at 1%

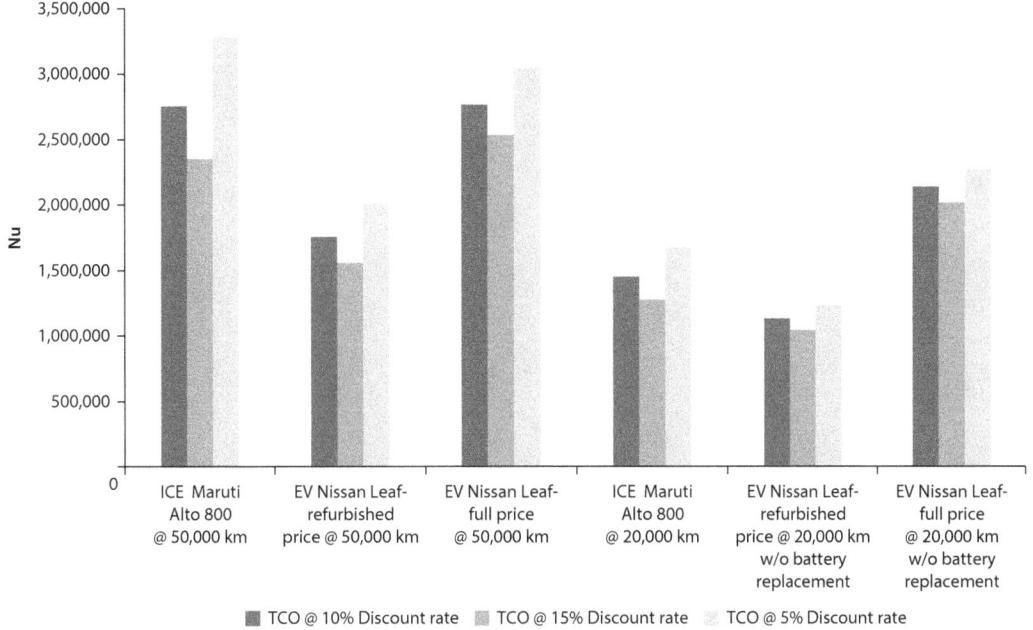

Source: World Bank analysis.
Note: EV = electric vehicle; ICE = internal combustion engine; TCO = total cost of ownership.

unique characteristics of taxi operations that differ from those for private vehicles. Other factors include, for example, the taxi drivers' ability to afford a higher upfront cost, access to vehicle financing, need for fast charging during the day, parking availability in public housing areas, revenue impact, and marketing opportunities. More information will be needed to develop the incentives program/strategy for taxis.

For the purpose of the aggregate analysis in this report, it is assumed that no incentives are required for taxis to meet the target set in each scenario given the stronger financial case for switching.

Results of TCO Analysis for Government Fleet

The government also has a plan to replace its fleet with EVs. For the purpose of the aggregate analysis in this report, the TCO analysis on private vehicles is assumed to be applicable to the government fleet. Further information on government vehicle use characteristics will be needed to conduct a more detailed TCO analysis.

APPENDIX D

Suppliers of CHAdeMO (CCS/AC) Fast Charging Equipment

A broad range of suppliers is available for the supply of fast charging equipment (table D.1). Differences among suppliers relate to the charging standard (CHAdeMO, Combined Charging Standard [CCS], alternating current [AC], GB/T, or Tesla), charging power (semifast to superfast), the number of outlets, and certification (CE, UL, or TUV).

More information about CHAdeMO suppliers can be found at: http://www.chademo.com/wp/chademocharger.

Table D.1 List of Suppliers for CHAdeMO Fast Charging Equipment

Supplier	Type	Standards	Power	Estimated cost[a]	Compatible cars
ABB (Switzerland)	Terra	CHAdeMO	50 kW	US$24,000	Nissan, Mitsubishi, Tesla (adaptor)
	Terra C	CCS, CHAdeMO	50 kW	US$40,000	BMW, Volkswagen, GM, Porsche, Audi, Nissan, Mitsubishi, Citroen, Tesla (adaptor), Daimler
	Terra	CCS, CHAdeMO, AC	50 kW	US$47,000	BMW, Volkswagen, GM, Porsche, Audi, Nissan, Mitsubishi, Citroen, Renault, Tesla (adaptor), Daimler
ICU (Netherlands)	QC	CHAdeMO	50 kW	US$40,000	Nissan, Mitsubishi, Tesla (adaptor)
Fuji Electric (Japan)	FRCA25C	CHAdeMO	25 kW	US$33,000	Nissan, Mitsubishi, Tesla (adaptor)
Eaton (United States)	Eaton/Takoaka	CHAdeMO	50 kW		Nissan, Mitsubishi, Tesla (adaptor)
Signet Systems (Korea, Rep.)	HB50K-EV	CHAdeMO	50 kW		Nissan, Mitsubishi, Tesla (adaptor)
PNE Solutions (Korea, Rep.)	PNE	CHAdeMO	50 kW		Nissan, Mitsubishi, Tesla (adaptor)
Delta Electronics (Taiwan, China)	Quick charger	CHAdeMO	50 KW		Nissan, Mitsubishi, Tesla (adaptor)
DBT (France)	Nissan chargers	CHAdeMO	40 kW	US$15,000	Nissan, Mitsubishi, Tesla (adaptor)
Siemens (Germany)	Efacec	CHAdeMO, CCS, AC	50 kW		BMW, Volkswagen, GM, Porsche, Audi, Nissan, Mitsubishi, Citroen, Renault, Tesla (adaptor), and Daimler

Note: a. Prices are indicative; prices vary considerably per specification, location and volume.

APPENDIX E

Possible Location of Charging Stations in Thimphu

The City of Thimphu is organized in 12 neighborhoods. Possible fast charging locations, as identified by the City of Thimphu, are shown in map E.1.

Map E.1 Possible Charging Locations in Thimphu Thromde

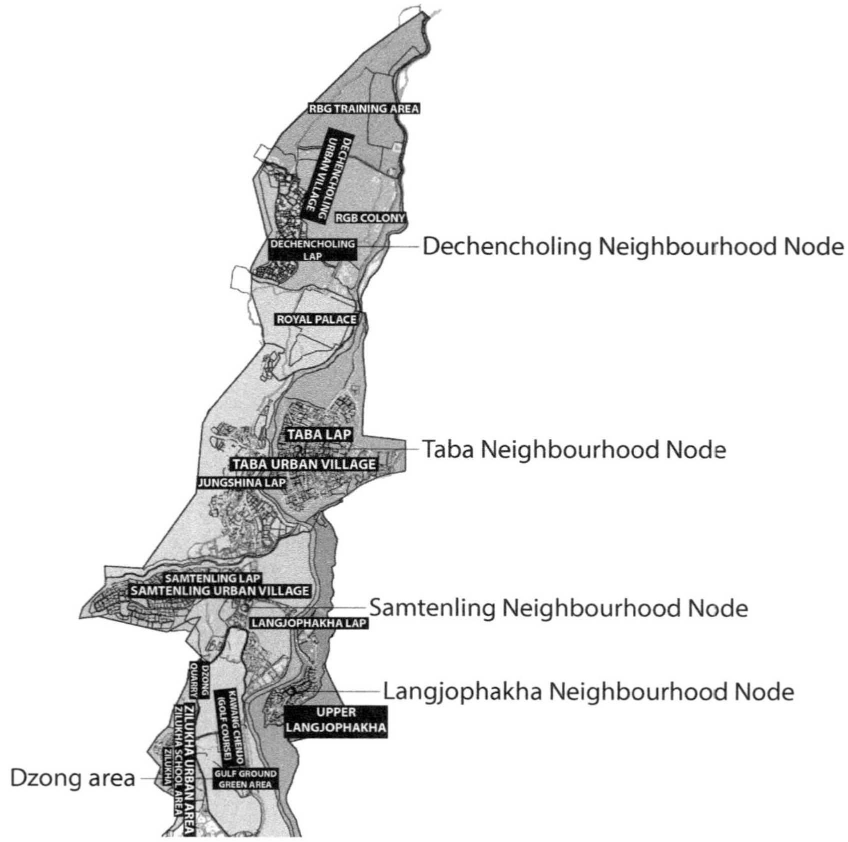

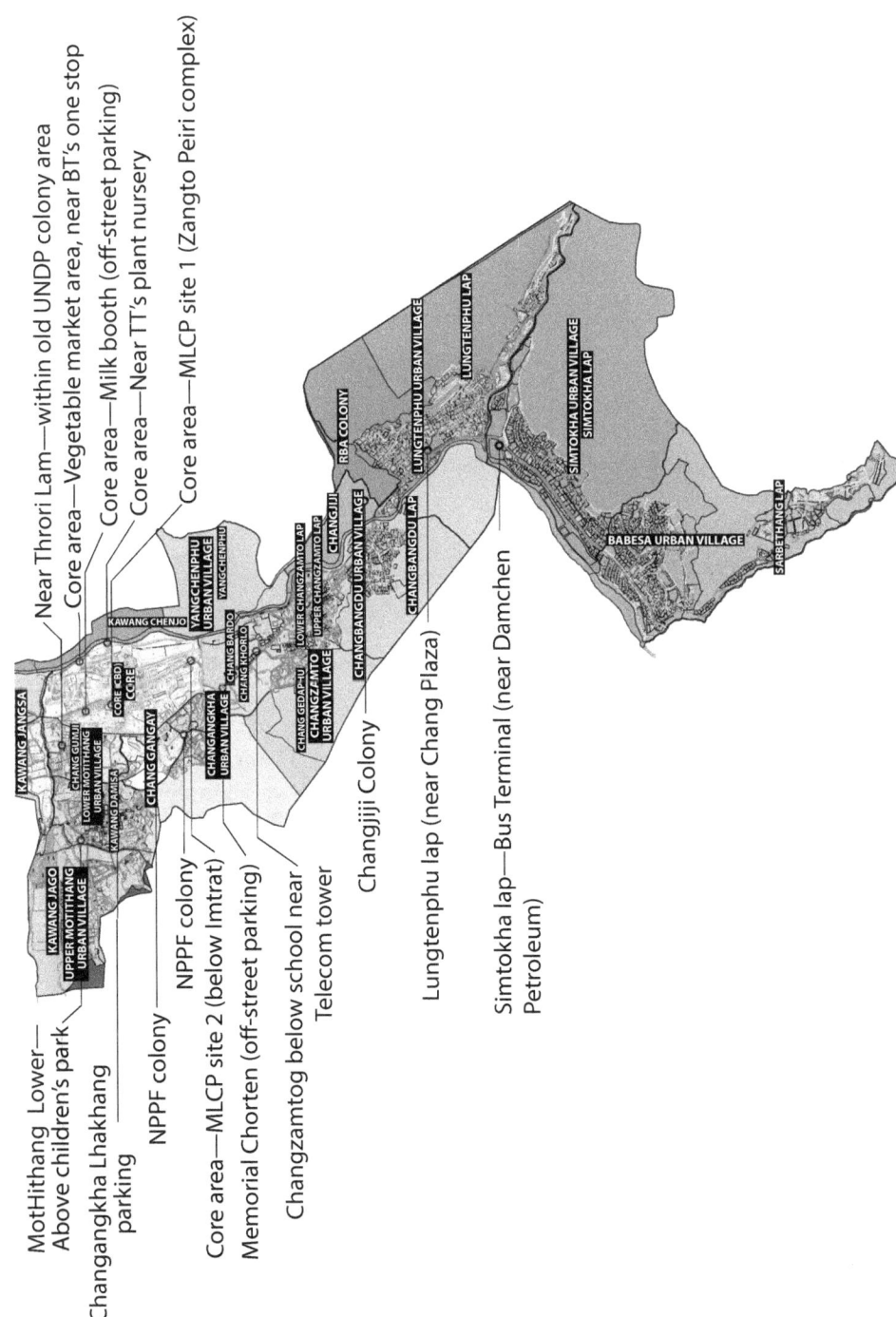

Source: Thimphu Thromde.

Initial assessment for charging locations include:

- Simtokha lap—bus terminal (near Damchen Petroleum)
- Lungtenphu lap (near Chang Plaza)
- Changjiji Colony (has ample parking facilities where multiple connection charging stations can be installed)
- Changzamtog (below the school near the Telecom tower)
- Memorial Chorten (off-street parking)
- NPPF colony
- Core area—MLCP site 2 (below Imtrat)
- Core area—Changlam (opposite JoJo)
- Core area—MLCP site 1 (Zangtogpelri complex)
- Core area—milk booth (off-street parking)
- Core area—vegetable market area near BT's one-stop shop
- Core area—near TT's plant nursery
- Near Throri Lam—within old United Nations Development Programme colony area
- Changangkha Lhakhang parking
- Mothithang Lower—above children's park next to the bus stop, undeveloped piece of land on the main road
- Dzong area
- Samtenling Node
- Langjophakha
- Taba—an undeveloped dirt road location about 100 m from the main road, a possible location in 2015 when the dirt road is changed to asphalt and connected to the new bridge to the west side.
- Dechenchholing satellite town, roadside location with power connection close by. This is the northernmost location in Thimphu and could serve the area with about 10,000 people and villages above Thimphu with another 15,000 people.

APPENDIX F

Comparison of Bus Transport Technologies

International Experience and Best Practice

Around the world, numerous bus pilots are conducted using new and more sustainable transport techniques. Transport authorities often select these techniques for specific (and local) policy objectives. The main drivers are improvement of local air quality and CO_2 emissions reduction. The availability of infrastructure, resources, or expertise often plays an important role in the decision-making process. For example, in the city of Rotterdam (the Netherlands), pilot projects with hydrogen buses are being conducted because of a hydrogen production facility in the nearby harbor. Other cities and regions choose different techniques such as compressed natural gas (CNG), biofuels, or trolleys.

In Thimphu, the current buses adhere mostly to the Euro 3 norm. The Royal Government of Bhutan (RGoB) and City of Thimphu are investigating even further emissions restrictions or new techniques for buses. The change from existing to new techniques has consequences for three indicators: economic, environmental, and operational performance. To make a seamless switch in the running of operations, a thorough insight into these three factors is required. Because of this, such a switch is often preceded by a pilot project to verify the assumptions.

Before a new technique is implemented, the following three indicators need to be investigated and possibly verified in a pilot project:

1. *Economic performance:* Obtaining experience with costs of hardware, maintenance, infrastructure, and education of skilled personnel (such as mechanics and bus drivers).
2. *Environmental performance:* Measuring the impact on the environment (CO_2, nitrous oxide [NOx], and particulate matter [PM10]).

3. *Operational performance:* Obtain insight into operational differences with, for instance, planning of charging times and charging locations for electric buses. Also practical experience like driving behavior, real life range, and driver and passenger satisfaction can be obtained.

Obtaining experience with the technique in order to make a balanced decision about possible upscaling of buses in the future is required. Although lessons can be learned from other pilots, local differences (for example, route schedules, driving speed and driving style, number of passengers, and topography) are of such significant influence that local pilots are necessary. Pilot buses should be used in the "normal" time schedule (but less crucial routes) to see if they fit the requirements of the local transport provider. Pilot projects also give manufacturers the experience to prepare for large-scale production and implementation.

International experience with electric bus pilots includes those in China, the United Kingdom, and the United States:

- **Shenzhen, China.** In Shenzhen, 780 electric buses (from the company BYD, based in Shenzhen) are being deployed throughout the city region. The project started with a two-bus pilot of six months in 2011. After 6 months another 200 buses were added to the fleet, and in 2012 the final 580 units were delivered. The buses are operating on 44 bus lines, and about 100 charging stations are in service. The buses charge mainly overnight, using a charging cable. The average energy usage is about 1.3 kW/km. The seating capacity is 31 persons with an additional 37 standing capacity. Not only buses, but also a total of 300 BYD electric taxis, are in operation. Although little information is available, it is likely that both environmental issues and empowering domestic manufacturers are driving the boost of electric buses in this case. No information is available about possible government funding.

- **Milton Keynes, England.** In Milton Keynes, trials are being conducted with full electric buses. The eight buses are 9.4 meters long and have a seating capacity of 37 with an additional standing capacity for 9 persons. The buses have a full charge at night and then charge wirelessly (inductive) during the day. On-route charging allows the buses to have a smaller (expensive) battery pack. There are only two inductive charging points for all the buses on the 15-mile bus route. The pilot acquired £640,000 in government funding.

- **California, United States.** In the State of California, the public transport provider Foothill Transit has 15 electric "opportunity charging" buses in operation; the first buses were in operation in 2010. The buses charge with an overhead fast charger. The average charging time is between 5 and 10 minutes, which translates to a driving range of about 45–65 km. In 2010, Foothill Transit was awarded a US$10.17 million grant to purchase 12 buses and charging infrastructure.

Comparison of Bus Transport Technologies

Table F.1 Pros and Cons of Various Bus Technologies

Technique	Pros	Cons	Average range
Diesel Euro 3	Proven technology Local knowledge about technique, reparation and spare parts	Driving on fossil fuels Impact on the environment	300–600 km
Diesel Euro 5	Proven technology Local knowledge about technique, reparation and spare parts	Driving on fossil fuels Impact on the environment	300–600 km
Diesel Hybrid	Proven technology	Driving on fossil fuels Expensive technology in comparison to Euro 5, relatively lower impact on environment	300–600 km
Electric opportunity charging (OC)—overhead charging (example California)[a]	No air pollution No emission No noise pollution Relatively small battery pack is needed	High investment for infrastructure Less flexible with changing route schedules Limited driving range Only zero-emission when using renewable energy Limited local knowledge about technique, reparation, spare parts	40–100 km
Electric plug-in—charging with cable (example BYD Shenzhen) (Destination Charging [DC])[b]	No air pollution No emission No noise pollution	Large battery pack needed High costs for battery Impact on passenger capacity Limited amount of manufactures Only zero-emission when using renewable energy Limited local knowledge about technique, reparation, spare parts High investments for infrastructure Operational limitations due to charging time	100–250 km

Source: World Bank analysis.
a. Opportunity charging (OC) is a technique for full electric buses whereby charging is done on route at bus stops. This charging can be with overhead fast charging or induction charging.
b. Destination charging (DC) is a technique for full electric buses whereby charging is done at the bus station when the bus in not in service.

In addition to electric vehicles, other bus technologies also can make transportation more sustainable. Table F.1 summarizes the advantages and disadvantages of various available technologies.

Bus Transport Total Cost of Ownership

The total cost of ownership (TCO) of buses depends on many variables, and TCO calculations should involve all stakeholders, such as the public transport authority, vehicle manufacturers, infrastructure providers, transport companies, and local authorities, to obtain specific and reliable output. In this TCO calculation, a brief analysis is used to get a first rough insight into the costs of some of the available technologies, with most of the critical assumptions taken into account. Other important factors, however, such as route length, terrain type, speed, use of air-conditioning, and use of staff, are not included in the model. Additional analyses and calculations will be needed to get a more detailed insight into the costs.

TCO Assumptions

For the TCO analysis, the following general assumptions were applied:

- All buses transition to other techniques at the same time (no in-growth scenarios, and possible impacts of specific bus lines are not taken into account).
- Prices for buses and infrastructure are based on a European price level.
- Only 12-meter buses.
- An average of 45,000 kilometers per year per bus.
- Calculations are based on 32 buses (10 lines) for public transport currently in operation.
- Depreciation time of 10 years for all buses.
- CO_2 emissions price of US$30 per ton of CO_2; this is incorporated into the TCO.

Additional assumptions about energy usage, fuel price, price per bus, and maintenance costs for the different techniques are listed in table F.2. The table also lists the results of the analysis, in terms of the TCO for 10 years. The price per kilometer for the various techniques is shown in figure F.1.

Table F.2 Bus Transport TCO Assumptions and Results

Technique	Assumptions	Total cost for 10 years (million US$)	Price increase
Diesel Euro 3	Energy usage: 0.6 l/km Fuel price: US$0.89 per liter Price per bus: US$150,000 Maintenance costs per km: US$0.19	16.1	0% (reference)
Diesel Euro 5	Energy usage: 0.45 l/km Fuel price: US$0.89 per liter Price of bus: US$250,000 Maintenance costs per km: US$0.18	17.2	+7%
Diesel Hybrid	Energy usage: 0.32 l/km Fuel price: US$0.89 per liter Price of bus: US$300,000 Maintenance costs per km: US$0.20	17.7	+10%
Electric opportunity charging	Energy usage: 1.3 kW/km Fuel price: US$0.045 per kWh Price of bus: US$450,000 Maintenance costs per km: US$0.09 Price of charger: US$200,000 Number of chargers: 20 (10 bus lines, 2 per line)	23.2	+44%
Electric plug-in	Energy usage: 1.3 kW/km Fuel price: US$0,045 per kWh Price of bus: US$450,000 Maintenance costs per km: US$0.09 Price of charger: US$10,000 Number of chargers: 16 (double socket)	21.4	+33%

Source: World Bank analysis.

Figure F.1 Bus Transport Total Cost of Ownership per Kilometer for Different Technologies

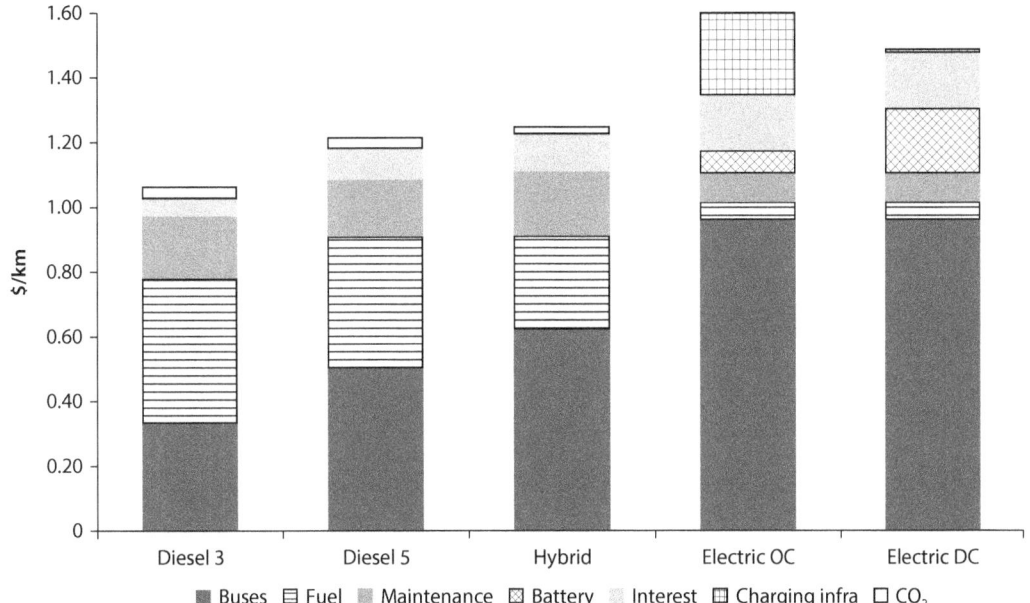

Source: World Bank analysis.
Note: A CO_2 emission price of US$30 per ton CO_2 is incorporated in the TCO. DC = destination charging; OC = opportunity charging; TCO = total cost of ownership.

From the outcomes of the model, it is clear that newer sustainable techniques have a significant price tag. The more expensive buses and higher costs for charging infrastructure are the main causes for the increase of the price. However, the price for maintenance costs and energy is lower for electric buses in comparison with traditional diesel buses. Also, the general assumption is that electric buses last longer than traditional diesel buses (15 years vs. 10 years) because they have a simpler mechanical system with fewer (moving) parts. This last assumption is not taken into account in these calculations.

Life Cycle Analysis Carbon Dioxide Emissions

The different technologies for buses come with different emissions profiles. For a first impression of environmental performance, a (simple) life cycle analysis (LCA) is performed for the technologies under consideration. The LCA looks at CO_2 equivalent emissions related to the production and operation of the system, including the hardware of the bus, fuel, charging infrastructure, and battery. The emissions related to the production of the diesel engines, the production of the electric transmission systems, and the "end of life" emissions have not been taken into account.[1]

Table F.3 shows the assumptions of emissions used for the LCA. Results of the analysis, in annual CO_2 equivalent emissions per bus, are also listed in the table and shown in figure F.2. As shown in the figure, electric buses have a significant

Table F.3 Life Cycle Analysis Assumptions and Results

Technology	Emission assumptions (CO_2 kg/km)	Yearly CO_2 equivalent emissions per bus (CO_2 kg/year*bus)
Diesel Euro 3	Hardware: 0.28 WTW fuel: 1.2	66,600
Diesel Euro 5	Hardware: 0.28 WTW fuel: 1.1	62,100
Diesel Hybrid	Hardware: 0.28 WTW fuel: 0.75 Battery: 0.05	46,350
Electric opportunity charging (OC)	Hardware: 0.28 WTW fuel: 0 Charging infrastructure: 0.05 Battery: 0.2	23,850
Electric plug-in (DC)	Hardware: 0.28 WTW fuel: 0 Charging infrastructure: 0.05 Battery: 0.2	23,850

Source: World Bank analysis.
Note: DC = destination charging; OC = opportunity charging; WTW = Well-to-wheel.

Figure F.2 Life Cycle Analysis Results: Annual CO_2 Emissions per Bus for Different Technologies

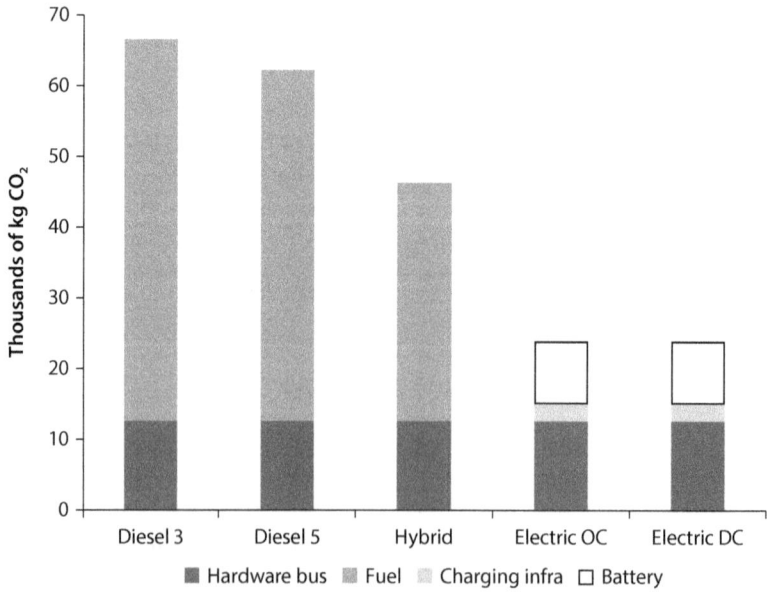

Source: World Bank analysis.
Note: Hardware bus: includes CO_2 equivalent emissions connected to the construction of the generic bus components (materials used and fabrication); Fuel: the CO_2 equivalent emissions related to the production and use of WTW (well-to-wheel) fuel; Charging infra: CO_2 equivalent emissions related to the materials used and production of the charging infrastructure; Battery: CO_2 equivalent emissions related to a battery pack for an electric bus. LCA = life cycle analysis.

impact on the reduction of CO_2 emissions. However, as noted in a recent World Bank report, the impact of 8–10 people taking a (diesel) bus instead of their private cars will have a comparable effect.[2]

Notes

1. In general, LCA research for bus technologies shows a wide variety of results. This is mainly because of the rapid development of technologies and the wide range in available technologies, types, usage, and production methods for each bus, battery, or charging infrastructure.
2. The report, "Bhutan: City Bus Service Technical Assistance Report" (World Bank 2015), points out that "A key fact to note from the analysis above is that getting 8–10 car drivers per day to take the bus instead of their private cars would yield roughly as much emissions reduction as 1 electric bus."

Environmental Benefits Statement

The World Bank Group is committed to reducing its environmental footprint. In support of this commitment, the Publishing and Knowledge Division leverages electronic publishing options and print-on-demand technology, which is located in regional hubs worldwide. Together, these initiatives enable print runs to be lowered and shipping distances decreased, resulting in reduced paper consumption, chemical use, greenhouse gas emissions, and waste.

The Publishing and Knowledge Division follows the recommended standards for paper use set by the Green Press Initiative. The majority of our books are printed on Forest Stewardship Council (FSC)–certified paper, with nearly all containing 50–100 percent recycled content. The recycled fiber in our book paper is either unbleached or bleached using totally chlorine-free (TCF), processed chlorine-free (PCF), or enhanced elemental chlorine-free (EECF) processes.

More information about the Bank's environmental philosophy can be found at http://www.worldbank.org/corporateresponsibility.

www.ingramcontent.com/pod-product-compliance
Lightning Source LLC
Chambersburg PA
CBHW060314240426
43661CB00059B/2757